Concurrent Engineering in Construction Projects

Also available from Taylor & Francis

Concurrent Engineering in Construction Projects

Edited by
Chimay J. Anumba,
John M. Kamara and
Anne-Francoise Cutting-Decelle

Routledge
Taylor & Francis Group

LONDON AND NEW YORK

First published 2007
by Taylor & Francis
2 Park Square, Milton Park, Abingdon, Oxon OX14 4RN

Simultaneously published in the USA and Canada
by Taylor & Francis
711 Third Avenue, New York, NY 10017

*Taylor & Francis is an imprint of the Taylor & Francis Group,
an informa business*

First issued in paperback 2016

© 2007 Chimay J. Anumba, John M. Kamara, Anne-Francoise
Cutting-Decelle, editorial selection; individual chapters, the contributors

Typeset in Sabon by
Newgen Imaging Systems (P) Ltd, Chennai, India

British Library Cataloguing in Publication Data
A catalogue record for this book is available
from the British Library

Library of Congress Cataloging in Publication Data
    Concurrent engineering in construction projects / edited by
Chimay J. Anumba, John M. Kamara, and Anne-Francoise
Cutting-Decelle.
        p. cm.
    Includes bibliographical references.
    1. Construction industry–Quality control. 2. Concurrent
engineering. I. Anumba, C. J. (Chimay J.) II. Kamara, John M.
III. Cutting-Decelle, Anne-Francoise.
    TH438.C638 2006
    690.068′5–dc22   2005028154

ISBN13: 978–0–415–39488–8 (hbk)
ISBN13: 978–1–138–97147–9 (pbk)

Contents

Figures

Tables

Contributors

Robert Amor is an Associate Professor, and Head of the Department of Computer Science at the University of Auckland, New Zealand.

Chimay J. Anumba is founding Director of the Centre for Innovative and Collaborative Engineering (CICE) and Professor of Construction Engineering and Informatics at Loughborough University.

Ghassan Aouad is Head of the School of Construction and Property Management and Director of the £3M EPSRC IMRC Centre (SCRI) at Salford University.

Andrew N. Baldwin is the Dean of the Faculty of Construction and Land Use at the Hong Kong Polytechnic University.

Nasreddine (Dino) M. Bouchlaghem is a Professor of Architectural Engineering at Loughborough University, UK.

Patricia M. Carrillo is Professor of Strategic Management in Construction at Loughborough University, UK.

Mike Clift is an Associate Director in the Centre for Whole Life Performance at BRE undertaking whole life cost and building performance work for a range of clients.

Rachel Cooper is a Professor, at Lancaster University (UK) and undertakes research in the areas of design and construction process, new product development, design management and socially responsible design.

Anne-Francoise Cutting-Decelle is currently Professor in Advanced Manufacturing Technologies at the University of Evry, France (IUT-Department of Organisation and Production Engineering), and Academic Visitor at Loughborough University (UK).

Alistair K. Duke is Principal Research Engineer at BT Research and Venturing, Ipswich, UK.

Martin Fischer is Professor of Civil and Environmental Engineering and (by Courtesy) Computer Science, at Stanford University. He is also Director, Center for Integrated Facility Engineering (CIFE) at the Stanford University, USA; and Founder/Chief Scientist, Common Point, Inc.

Andrew Fleming is a Research Fellow at Salford University, UK.

Michail (Mike) Kagioglou is a professor at the School of Construction and Property Management at Salford University, UK.

John M. Kamara is Senior Lecturer, School of Architecture, Planning and Landscape, Newcastle University, UK.

Malik M. A. Khalfan is working with Salford Centre for Research and Innovation at the Salford University as a Research Fellow.

Lauri Koskela is a Professor at Salford University, UK. He is a founding member of the International Group for Lean Construction.

Angela Lee is a Lecturer in the School of Construction and Property Management at Salford University, UK.

Line C. Pouchard is a Researcher in the Department of Computer Science and Mathematics at the Oak Ridge National Laboratory, USA.

Raimer J. Scherer is a Professor and Director of the Institute of Construction Informatics at the Technical University of Dresden, Germany.

Sheryl Staub-French is an Assistant Professor of Civil Engineering and Coordinator of the Engineering Management Program at the University of British Columbia, Vancouver, Canada.

Žiga Turk is Professor and Chair of Construction Informatics at the Faculty of Civil Engineering, University of Ljubljana, Slovenia.

Peter Walker is a Director in the Newcastle office of DEWJOC Architects, a practice with offices in the North East and London, UK.

Song Wu is a Research Fellow in the School of Construction and Property Management at the Salford University, UK.

Acknowledgements

We are grateful to all contributors who have spent a considerable amount of time putting together their chapters. We acknowledge the contribution of the various agencies and organisations that funded the research projects and initiatives on which this book is based. Mrs Jo Brewin (Professor Anumba's PA) played a major role in collating the chapters. We are also indebted to our families, whose continued love and support make ventures of this nature both possible and worthwhile.

The following organisations and publishers granted permission for the re-use of some of the illustrations in this book:

- BAA plc
- The Institution of Structural Engineers
- Elsevier
- Thomas Telford
- Emerald
- IBM Lotus Notes Screen Captures or other materials Copyright IBM Corporation. Used with permission of IBM Corporation. IBM, Lotus and Notes are trademarks of IBM Corporation, in the United States, other countries, or both.
- Microsoft product screen shots reprinted with permission from Microsoft Corporation.

This book is an output of the CIB (International Council for Research and Innovation in Building and Construction) Task Group TG33 on Collaborative Engineering.

Professor Chimay J. Anumba
Dr John M. Kamara
Professor Anne-Francoise Cutting-Decelle
(Editors)
July 2006

Chapter 1

Introduction to Concurrent Engineering in construction

John M. Kamara, Chimay J. Anumba and Anne-Francoise Cutting-Decelle

1.1 Introduction

The term Concurrent Engineering (CE) was coined in the late 1980s to explain the systematic method of concurrently designing both the product and its downstream production and support processes (Evbuomwan and Anumba, 1995; Huovila *et al.*, 1997). CE was proposed as a means to minimise product development time (Prasad, 1996). This was necessitated by changes in: manufacturing techniques and methods, management of quality, market structure, increasing complexity of products and demands for high quality and accelerated deliveries at reduced costs. These changes resulted in a shift in corporate emphasis with the result that, the ability to rapidly react to changing market needs and time-to-market became critical measures of business performance (Constable, 1994; Thamhain, 1994).

The earliest definition of CE by Winner *et al.* (1988) refers to 'integrated, concurrent design of products and their related processes, including manufacture and support' with the ultimate goal of customer satisfaction through the reduction of cost and time-to-market, and the improvement of product quality. CE embodies two key principles: integration and concurrency. Integration here is in relation to the process and content of information and knowledge, between and within project stages, and of all technologies and tools used in the product development process. Integrated concurrent design also involves upfront requirements analysis by multidisciplinary teams and early consideration of all lifecycle issues affecting a product. Concurrency is determined by the way tasks are scheduled and the interactions between different actors (people and tools) in the product development process. Table 1.1 shows a matrix of concurrency which can be used to assess the level of 'concurrency' within a project team (Prasad *et al.*, 1993).

The rows represent modes of operation and the columns represent the possible work-group configurations. A cooperating user is *a person who completes the work left unfinished by previous users* (Prasad *et al.*, 1993). Simultaneous users refer to other members of the project team who may access *the same design, tool or application concurrently or... different versions of product information tools or applications (PITA) at the same*

Table 1.1 Matrix of concurrency

No.	Modes of interactions	Work-group configurations		Simultaneous users	
		Single user	Cooperating users	Different versions	Same version
1	Access own products' interaction tools or applications (PITA)	Sequential engineering (SE)	SE	SE	SE
2	Run against their own data	SE	SE	SE	SE/CE
3	Access PITA belonging to other work-groups	SE/CE	CE	CE	CE
4	Access data belonging to other work-groups	CE	CE	CE	CE
5	Access both PITA and data from other work-groups	CE	CE	CE	CE

Source: Prasad et al. (1993).

time (Prasad *et al.*, 1993). The level of concurrency depends on the type of interactions, and this increases as one moves from top to bottom and from left to right (Table 1.1). It is observed that some situations are described as both sequential and concurrent: when simultaneous users run their own data, and when a single user accesses the PITA belonging to other work groups (Table 1.1). The interaction will be sequential if two or more users cannot edit and save changes to a document until another user has finished with it, even though they can be working in parallel.

The key features of CE can therefore be summarised to include the following:

- Concurrent and parallel scheduling of all activities and tasks as much as possible.
- Integration of product, process and commercial information over the lifecycle of a project; and integration of lifecycle issues during project definition (design).
- Integration of the supply chain involved in delivering the project through effective collaboration, communication and coordination.
- Integration of all technologies and tools utilised in the project development process (e.g. through interoperability).

1.1.1 Implementation of CE

CE is a philosophy which contains (or is implemented by) several methodologies. The attainment of 'integrated, concurrent design' requires a variety of

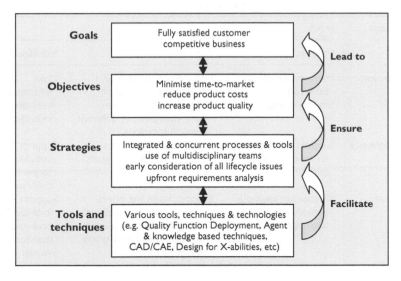

Figure 1.1 A framework for understanding CE.

Source: Kamara *et al*. (2000).

enablers which include tools (software applications), techniques, technologies and support structures. These enablers can be generic and can be used to support other concepts. The extent to which these principles are implemented determines the level to which the objectives of CE (e.g. shorter lead times) are realised. Figure 1.1 shows a framework for the implementation of CE with respect to the interrelationships between the goals, objectives, strategies and tactics (tools and technologies) for CE.

The goal of fully satisfying the customer and operating a competitive business, is made possible by shorter lead times (time-to-market), lower costs and high quality products. These in turn, arc achieved through rigorous requirements analysis, early consideration of all life-cycle issues affecting a product, integrated and concurrent product development and the use of multi-disciplinary teams and other strategies. The overall CE framework is facilitated by various tools and techniques which include: Quality Function Deployment (QFD), agent and knowledge based tools, computer aided design (CAD) and Computer Aided Manufacturing (CAM) tools, and other relevant tools (Evbuomwan and Sivaloganathan, 1994; Prasad, 1998).

CE enablers (tools and techniques) can be grouped into two broad categories which are interrelated: organisational and technological. Organisational enablers provide the framework for people and machines to work 'concurrently'. This includes: facilitating the work of multi-disciplinary

Table 1.2 Support requirement matrix for CE

Dimensions	Levels		
	Organisational	*Team*	*Individual*
Distribution	Move information between multiple sites	Reduce remoteness and promote exchange of information between team members at different physical locations	Make information available to individuals
Heterogeneity	Support organisations to achieve different missions	Support project teams to achieve different goals	Support individuals to perform different jobs
Autonomy	Discourage multiple individual stores of information	Support team members to work as individuals, or as a group, and transitions between these two types of working	Support individual's preferred manner of working

Source: Harding and Popplewell (1996).

teams, involving all relevant parties in the product development process, and managerial/technological support for organisational, team and individual levels of working. Table 1.2 summarises the kind of support required for CE at the organisational, team and individual levels with respect to distribution, heterogeneity and autonomy.

Technological enablers facilitate concurrent working within organisations. They include all the Information and Communications Technologies (ICTs) and computer-based applications required for integration, concurrent working, communication and collaboration.

1.1.2 Benefits of CE

The benefits of CE derive from the fact that it is focused on the design phase (Koskela, 1992) which determines and largely influences the overall cost of a product: as much as 80 per cent of the production cost of a product can be committed at the design stage (Dowlatshahi, 1994): Addressing all life-cycle issues up-front in the design stage and ensuring that the design is 'right-first-time' should therefore lead to cost savings, products that precisely match customers' needs, and which are of a high quality. The adoption of CE can also result in reductions in product development time of up to 70 per cent (Madan, 1993; Carter, 1994; Constable, 1994; Dowlatshahi, 1994; Evbuomwan *et al.*, 1994; Frank, 1994; Nicholas, 1994; Thamhain, 1994; Smith *et al.*, 1995; Prasad, 1996).

1.2   CE in construction

The success of CE in manufacturing, which is due to the benefits arising from its use, is one of the main motivations for adopting CE in construction (de la Garza *et al.*, 1994; Anumba and Evbuomwan, 1995; Evbuomwan and Anumba, 1995, 1996; Huovila and Serén, 1995; Hannus *et al.*, 1997; Kamara *et al.*, 1997; Love and Gunasekaran, 1997; Anumba *et al.*, 1999, etc.). It is also based on the assumption that because construction can be considered as a manufacturing process, concepts which have been successful in the manufacturing industry can bring about similar improvements in the construction industry. Furthermore, the goals and objectives of CE directly address the challenges that currently face the construction industry.

1.2.1   Construction as a manufacturing process

The interest in modelling construction as a manufacturing process is primarily based on the similarities between the two industries, and the assumption that, aligning the business processes of the construction industry to those of the manufacturing industry will significantly improve its competitiveness (Sanvido and Medeiros, 1990; Anumba and Evbuomwan, 1995; Anumba *et al.*, 1995; Crowley, 1996; Egan, 1998). Both the manufacturing and construction industries:

- produce engineered products that provide a service to the user;
- are involved in the processing of raw materials and the assembly of many diverse pre-manufactured components in the final products;
- utilise repeated processes in the design and production of their products;
- experience similar problems such as: the high cost of correcting design errors due to late changes, poor resource utilisation, and inadequate information management.

The differences between manufacturing and construction with regard to the location of production activities, and the production of 'one-off' facilities in construction, as opposed to mass production in manufacturing, have led to suggestions that the two industries are profoundly different (Sanvido and Medeiros, 1990; Crowley, 1996; Egan, 1998). However, the parallel between construction and manufacturing is not with respect to repeated (or mass-produced) *products*, but rather to the repeated *processes* that are involved in the design and production of products in both industries. The implication of this is that, developments in manufacturing such as CE which have led to improvements in productivity (as a result of *process* re-engineering) can be used in construction.

Table 1.3 The rationale for adopting CE in construction

Need for change in construction	Goals and principles of CE
The need for change in construction is brought about by the uncompetitive nature of the industry, and the inability to fully satisfy its clients with respect to costs, time and value	The goals and objectives of CE (Figure 1.1) include: customer satisfaction, competitive business, reduction of product development time and cost, improvement of quality and value
Integration of the construction process is seen as one of most important strategies to improve the notoriously fragmented construction industry	The use of CE facilitates the integration of the members of the product development team, and the manufacturing process, thereby improving the product development process
Emerging strategies for improving the construction process are inadequate; they only address one aspect of the problem, resulting in 'islands of automation' as in the case of computer-integrated construction strategies	As an amalgam of other methodologies, tools and techniques, CE provides a framework for not only integrating the construction process, but also the various tools and technologies that are used in the process

1.2.2   The relevance of CE principles to construction

Another justification for the adoption of CE in construction is based on the fact that the goals and strategies (principles) of CE directly address the problems in the construction industry. Table 1.3 provides a summary of how the needs in construction, discussed earlier, can be addressed by CE. This pairing of needs versus capabilities in support of CE in construction is further buttressed by the fact that, existing practices in the construction industry, which are similar to CE, can facilitate its successful implementation in construction.

It is therefore evident that CE has considerable potential in construction. Its capacity to provide an effective framework for integrating and improving the construction process is now also widely acknowledged in the industry (Anumba and Evbuomwan, 1997; Egan, 1998). From both the context in which it evolved (manufacturing), and its inherent features, CE can be matched to the construction process. Its implementation however, needs to suit the particular needs of the construction industry.

1.2.3   Implementation of CE in construction

The construction industry (otherwise referred to as the Architecture, Engineering and construction – AEC industry) is organised around projects that are paid for by clients who are technically not part of the industry. Construction projects are also delivered by many firms, unlike the manufacturing industry, where a greater proportion of the skills required may be held within one organisation. Achieving 'true concurrency' in AEC (Table 1.2), for example, might require users from one firm (e.g. structural engineering consultants) to access both PITA (product interaction tools and applications – for example, CAD workstations) and data from other work groups that might be located other firms.

Because of the project nature of the AEC industry, CE implementation in construction should be considered at both the project and organisational (i.e. individual consulting/contracting firms) levels. At the organisational level, it is relatively easier to devise strategies that reflect the requirements set out in Tables 1.1 and 1.2, which are somehow based on a single-organisation model. At the project level, 'concurrency' and 'integration' should focus only on issues pertaining to the project. The matrix of concurrency in Table 1.1 is also applicable at this level, but relatively more difficult to implement; some aspects of Table 1.2 (e.g. organisational, team and individual support for heterogeneity) may not be applicable since a specific project can be considered as a homogenous entity.

Other challenges for CE in AEC include the linkages between organisational (i.e. firm level) support structures and project level support requirements. Somebody operating at the organisational level may store data on different projects in their PITA; access to information relating to a specific project by somebody outside the organisation therefore becomes problematic. Another challenge relates to the role of clients who dictate the nature and form of the project organisation (through procurement and contractual strategies adopted), and in some cases, even the range of technologies that can be used. The fact that the project and organisational levels are influenced by different (and sometimes) opposing forces (i.e. client and industry), poses challenges for the linkages between the two.

1.3   Scope and outline of the book

This book is a collection of papers that reflect various research efforts on the implementation of CE in construction projects. The aim is to present the key issues and technologies important for the adoption of CE, starting with fundamental concepts and then going on to the role of organisational enablers and advanced information and communication technologies. The twelve chapters in the book are broadly divided into four main sections: introduction and foundations, organisational enablers, technological enablers and conclusions and future directions.

1.3.1   Introduction and foundations
(Chapters 1 and 2)

Chapter 1 provides a brief introduction into the principles and implementation issues for CE in construction. In Chapter 2, the theoretical foundations of CE are discussed. It is argued that CE is based on a new conceptualisation of engineering operations as transformation, flow and value generation. In the flow view, the basic thrust is to eliminate waste from flow processes. Thus, such practices as rework reduction, team approach, the release of information in smaller batches to following tasks, are promoted. In the value generation view, the basic thrust is to achieve the best possible value from the customer's point of view. Such practices as rigorous requirements analysis, systematised management of requirements during engineering and rapid iterations for improvement, are promoted.

1.3.2   Organisational enablers (Chapters 3 to 6)

Chapters 3 to 6 broadly address 'organisational enablers' for CE with respect to processes, procurement, client and organisational readiness for CE. Chapter 3 focuses on organisational 'readiness' to adopt (or be part of) a CE project process. It describes and discusses the development and implementation of a new readiness assessment model (the BEACON model) for the construction industry. Chapter 4 considers the role of the client in the implementation of CE in construction and describes an approach for incorporating the 'voice of the client' in construction projects that can facilitate both a CE approach and a focus on the needs of the client. In Chapter 5, procurement and contractual arrangements and their role in facilitating a CE approach, are discussed. In Chapter 6, the focus is on standardised processes. The authors provide an introduction to process management in construction and present the key principles of an improved holistic process for CE in construction that allows continuous learning through feedback mechanisms.

1.3.3   Technological enablers (Chapters 7 to 11)

Chapters 7 to 11 address 'technological enablers' for CE. Chapter 7 discusses ontologies and standards based approaches to interoperability, which provides a solid basis for the integration of tools and technologies within a CE framework. In Chapter 8, the integration of product and process representation of design and construction information is considered. An integrated product and process model is proposed and the chapter further defines some fundamental bases on which CE concepts can be developed. Chapter 9 looks at document management and shows how the proper management of documents not only provides information about all aspects of a project, but also facilitates the effective coordination of project activities and processes within a CE framework. Chapter 10 discusses the use of 4D models for effective coordination and project planning during the pre-construction phase of a project, and illustrates this with a case study. A Telepresence environment for distributed collaboration is the focus of Chapter 11. Chapter 12 explores various support tools for users within a CE environment that were developed through an EU project on Integrated Services and Tools for CE (ISTforCE).

1.3.4 Conclusions and future directions (Chapter 13)

Chapter 13 summarises the issues discussed in the book. It also uses the results of a study among North American academics and professionals on issues and challenges for CE implementation in construction to highlight future directions and the role of academics and industry practitioners in the adoption of CE on construction projects.

1.4 References

Anumba, C. J. and Evbuomwan, N. F. O. (1995), 'The Manufacture of Constructed Facilities', *Proceedings, ICAM '95*, pp. A.9.1–8.

Anumba, C. J. and Evbuomwan, N. F. O. (1997), 'Concurrent Engineering in Design-Build Projects', *Construction Management and Economics*, Vol. 15, No. 3, pp. 271–281.

Anumba, C. J., Evbuomwan, N. F. O. and Sarkodie-Gyan, T. (1995), 'An Approach to Modelling Construction as a Competitive Manufacturing Process', *12th Conference of the Irish Manufacturing Committee*, Cork, pp. 1069–1076.

Anumba, C. J., Baldwin, A. N., Bouchlaghem, N. M., Prasad, B., Cutting-Decelle, A. F., Dufau, J. and Mommessin, M. (1999), 'Methodology for Integrating Concurrent Engineering Concepts in a Steelwork Construction Project', *Advances in Concurrent Engineering, Proceedings of the CE99 Conference*, Bath, September 1999, Technomic Publishing Company, ISBN: 1 56676 790 3.

Carter, D. E. (1994), 'Concurrent Engineering', *Proceedings of the 2nd. International Conference on Concurrent Engineering and Electronic Design Automation (CEEDA '94)*, 7–8 April, pp. 5–7.

Constable, G. (1994), 'Concurrent Engineering – its Procedures and Pitfalls', *Measurement and Control*, Vol. 27, No. 8, pp. 245–247.

Crowley, A. (1996), 'Construction as a Manufacturing Process', in Kumar, B. and Retik, A. (eds), *Information Representation and Delivery in Civil and Structural Engineering Design*, Civil-Comp Press, Edinburgh, Scotland, pp. 85–91.

de la Garza, J. M., Alcantara, P., Kapoor, M. and Ramesh, P. S. (1994), 'Value of Concurrent Engineering for A/E/C Industry', *Journal of Management in Engineering*, Vol. 10, No. 3, pp. 46–55.

Dowlatshahi, S. (1994), 'A Comparison of Approaches to Concurrent Engineering', *International Journal of Advanced Manufacturing Technology*, Vol. 9, pp. 106–113.

Egan, J. (1998), 'Rethinking Construction', *Report of the Construction Task Force on the Scope for Improving the Quality and Efficiency of UK Construction*, Department of the Environment, Transport and the Regions, London.

Evbuomwan, N. F. O. and Anumba, J. C. (1995), 'Concurrent Life-Cycle Design and Construction', in Topping, B. H. V. (ed.), *Developments in Computer Aided Design and Modelling for Civil Engineering*, Civil-Comp Press, Edinburgh, UK, pp. 93–102.

Evbuomwan, N. F. O. and Anumba, J. C. (1996), 'Towards a Concurrent Engineering Model for Design-and-Build Projects', *The Structural Engineer*, Vol. 74, No. 5, pp. 73–78.

Evbuomwan, N. F. O. and Sivaloganathan, S. (1994), 'The Nature, Classification and Management of Tools and Resources for Concurrent Engineering', in Paul, A. J. and Sobolewski, M. (eds), *Proceedings of Concurrent Engineering: Research and Applications, 1994 Conference*, 29–31 August, Pittsburgh, Pennsylvania, PA, pp. 119–128.

Evbuomwan, N. F. O., Sivaloganathan, S. and Jebb, A. (1994), 'A State of the Art Report on Concurrent Engineering', in Paul, A. J. and Sobolewski, M. (eds), *Proceedings of Concurrent Engineering: Research and Applications, 1994 Conference*, 29–31 August, Pittsburg, Pennsylvania, PA, pp. 35–44.

Frank, D. N. (1994), 'Concurrent Engineering: A Building Block for TQM', *Annual International Conference Proceedings – American Production and Inventory Control Society (APICS)*, Falls Church, VA, USA, pp. 132–134.

Hannus, M., Huovila, P., Lahdenpera, P., Laurikka, P. and Serén, K.-J. (1997), 'Methodologies for Systematic Improving of Construction Processes', *Concurrent Engineering in Construction: Proceedings of the First International Conference*, in Anumba, C. J. and Evbuomwan, N. F. O. (eds), Institution of Structural Engineers, London, pp. 55–64.

Harding, J. A. and Popplewell, K. (1996), 'Driving Concurrency in a Distributed Concurrent Engineering Project Team: A Specification for an Engineering Moderator', *International Journal of Production Research*, Vol. 36, No. 3, pp. 841–861.

Huovila, P. and Serén, K.-J. (1995), 'Customer-Oriented Design Methods for Construction Projects', *International Conference on Engineering Design, ICED 95*, Praha, 22–24 August, pp. 444–449.

Huovila, P., Koskela, L. and Lautanala, M. (1997), 'Fast or Concurrent: The Art of Getting Construction Improved', in Alarcon, L. (ed.), *Lean Construction*, A. A. Balkema, Rotterdam, pp. 143–159.

Kamara, J. M., Anumba, C. J. and Evbuomwan, N. F. O. (1997), 'Considerations for the Effective Implementation of Concurrent Engineering in Construction', in Anumba, C. J. and Evbuomwan, N. F. O. (eds), *Concurrent Engineering in Construction: Proceedings of the First International Conference*, Institution of Structural Engineers, London, pp. 33–44.

Kamara, J. M., Anumba, C. J. and Evbuomwan, N. F. O. (2000), 'Developments in the Implementation of Concurrent Engineering in Construction', *International Journal of Computer-Integrated Design and Construction*, Vol. 2, No. 1, pp. 68–78.

Koskela, L. (1992), 'Application of the New Production Philosophy to Construction', *Technical Report No. 72*, Centre for Integrated Facility Engineering (CIFE), Stanford University, CA, USA, September, 1992.

Love, P. E. D. and Gunasekaran, A. (1997), 'Concurrent Engineering in the Construction Industry', *Concurrent Engineering: Research and Applications* Vol. 5, No. 2, pp. 155–162.

Madan, P. (1993), 'Concurrent Engineering and its Application in Turnkey Projects Management', *IEEE International Management Conference*, pp. 7–17.

Nicholas, J. M. (1994), 'Concurrent Engineering: Overcoming Obstacles to Teamwork', *Production and Inventory Management*, Vol. 35, No. 3, pp. 18–22.

Prasad, B. (1996), *Concurrent Engineering Fundamentals, Volume 1: Integrated Products and Process Organization*, Prentice Hall PTR, NJ.

Prasad, B. (1998), 'Editorial: How Tools and Techniques in Concurrent Engineering Contribute Towards Easing Co-operation, Creativity and Uncertainty', *Concurrent Engineering: Research and Applications*, Vol. 6, No. 1, pp. 2–6.

Prasad, B., Morenc, R. S. and Rangan, R. M. (1993), 'Information Management for Concurrent Engineering: Research Issues', *Concurrent Engineering: Research and Applications*, Vol. 1, No. 1, pp. 3–20.

Sanvido, V. E. and Medeiros, D. J. (1990), 'Applying Computer-Integrated Manufacturing Concepts to Construction', *Journal of Construction Engineering and Management*, Vol. 116, No. 2, pp. 365–379.

Smith, J., Tomasek, R., Jin, M. G. and Wang, P. K. U. (1995), 'Integrated System Simulation', *Printed Circuit Design*, Vol. 12, No. 2, pp. 19–20, 22, 63.

Thamhain, H. J. (1994), 'Concurrent Engineering: Criteria for Effective Implementation', *Industrial Management*, Vol. 36, No. 6, pp. 29–32.

Winner, R. I., Pennell, J. P., Bertrend, H. E. and Slusarczuk, M. M. G. (1988), 'The Role of Concurrent Engineering in Weapons System Acquisition', *IDA Report R-338*, Institute for Defence Analyses, Alexandria, VA.

Chapter 2

Foundations of Concurrent Engineering

Lauri Koskela

2.1 Introduction

The evolution of design management practice can conveniently be grouped into three periods: design as craft, sequential engineering and Concurrent Engineering (CE). Up to the Second World War, most industrial design was carried out by a small group of designers or a single generalist designer.[1] The products were simpler; the production processes were simpler. Thus, there were no major needs for systematised methods of design management and coordination. The period after the Second World War was characterised by the diffusion and further development of methods originated in wartime production of weapons. Also the development of large-scale systems such as telephone, television, etc. stimulated this evolution. Such approaches as systems engineering and project management grew out of these efforts. In established industries like car production, product development and design was organised in a roughly similar fashion to production: experts were grouped into different sections, departments, etc., and the design work flowed between these. The common feature was to organise design as a sequential realisation of design tasks. During the 1980s, the new concept of concurrent (or simultaneous) engineering emerged. In 1986, a report by the Institute for Defense Analyses coined the term CE to explain the systematic method of concurrently designing both the product and its downstream production and support processes. That report provided the first definition of CE as follows (Carter and Baker, 1992):

> Concurrent Engineering is a systematic approach to the integrated, concurrent design of products and their related processes, including manufacturing and support. This approach is intended to cause the developers, from the outset, to consider all elements of the product life cycle from concept through disposal, including quality, cost, schedule, and user requirements.

Today, CE is widely applied in practice, and it is also an increasingly popular research topic. However, a closer study of related case studies, reports and books shows that there is little agreement on the definition, basic features, and methods of CE. Thus, in an overview on CE (Prasad, 1996) not less than eight common definitions of CE are listed. A literature study shows that there are subsequent views on the basic nature of CE:

- CE equals teamwork. As Schrage (1993) states: 'Unfortunately, many companies believe they are implementing CE by convening multifunctional teams, which in reality is only one of 10 characteristics.'
- CE requires computerising. 'All characteristics (of CE) are dependent on a computing environment...' (Schrage, 1993).
- CE is a special approach to engineering. This view is exemplified by the recent Guide to the Project Management Body of Knowledge (Project Management Institute, 1996) with a very short discussion of Fast Tracking as the only reference to the ideas of CE. Thus, it is implied that there is a mainstream approach to engineering projects, and CE is a special approach, rarely needed.
- CE is a philosophy. 'Concurrent engineering is a philosophy and not a technology' (Jo et al., 1993).
- CE is a set of methods or tools. This 'recipe view' is common among the many authors giving 'how to' lists for CE implementation.
- CE is a Western attempt to understand Japanese product development practices (Tomiyama, 1995). After all, many, if not most, practices of CE have their origin in Japan (Sobek and Ward, 1996).

It is clear that further development and use of CE, for being successful, necessitates overcoming this confusion surrounding the topic. The basic argument of this presentation[2] is that traditionally, design and engineering has been viewed as transformation, whereas CE is based on mostly intuitive understanding of design and engineering as flow and value generation. Thus, basically CE is a conceptual and theoretical innovation. For defining and understanding CE, we have thus to focus on its theoretical foundations.

In a related stream of work (Koskela, 1992, 2000), it has been shown that the mentioned three conceptualisations explain the developments of production management in the twentieth century. These three concepts, transformation, flow and value generation, are not alternative, competing theories of production, but rather partial and complementary. What is needed is a production theory and related tools that fully integrate the transformation, flow and value concepts. As a first step towards this, we can conceptualise production simultaneously from these three points of view: transformation, flow and value. Such an integrated approach has been called the TFV theory of production. The following analysis is based on and extends this understanding of production operations.

2.2 Design as transformation

2.2.1 Sequential design

The ideas of scientific management soon diffused into the management of design. As a solution corresponding to the assembly line, serialising the design process, and determining a standard flow of design, was reached (Dasu and Eastman, 1994). Specialisation and associate division of work formed another part of the solution (Midler, 1996). These ideas seem to have guided design management in established industries, like car manufacturing, where product design is a recurrent activity.

However, the efforts to tackle large, unprecedented engineering projects in the war and in the 1950s stimulated new developments (Morris, 1994). One precursor was systems engineering,[3] which aimed at systematising large-scale system development (Hall, 1962). A generic flow of engineering tasks is one core issue of systems engineering (for a contemporary systems engineering methodology, see, for example, Methodik (1986)).

Another newcomer was project management. Morris describes the classic[4] – and still current – project management approach as follows (Morris, 1994):

> first, what needs to be done; second, who is going to do what; third, when actions are to be performed; fourth, how much is required to be spent in total, how much has been spent so far, and how much has still to be spent.... Central to this sequence is the Work Breakdown Structure (WBS).

According to Turner (1993), scope management is the raison d'être of project management. The purpose of scope management can be defined as follows: (1) an adequate or sufficient amount of work is done; (2) unnecessary work is not done; (3) the work that is done delivers the stated business purpose. The scope is defined through the work breakdown structure.

Thus, it is obvious that the project management discipline is a pure application of the transformation and its hierarchical decomposition. Also project management tools, like cost control and the Critical Path Method (CPM), are typically based on the transformation way of thinking (Koskela and Howell, 2002).

The transformation concept is also generally acknowledged in design science. Hubka and Eder (1988) state that:

> Engineering Design is a process...through which information in the form of requirements is converted into information in the form of description of technical systems.

In a similar vein[5] (Mistree *et al.*, 1993):

> Designing is a process of converting information that characterise the needs and requirements for a product into knowledge about a product.

Indeed, the conventional conceptualisation of design, in practice as well as in research, is based on the transformation model. In the framework of this conceptualisation, improvement of design and design management has become channelled – beyond tools for coordinating the whole design effort, as discussed earlier – into tools for enhancing the efficiency of individual tasks (CAD, calculation models, simulation models, decision support tools). The focus may be on decision making, with the premise that the principal content of design tasks is made up of decisions (Mistree *et al.*, 1993), or on problem solving (Murmann, 1994).

2.2.2 Anomalies

The identification of the problems caused by the prevailing organisation and management of product development and design started a search towards new methods in the 1980s. Putnam (1985) observed:

> Slow product launch, poor quality, and inefficiency are not isolated problems, nor are they symptoms of failure of individual functions of the business. The problems are related and reflect trouble in how those functions interact. The typical U.S. business links its design, manufacturing, and quality control departments only at points where a product moves from one department to the next. In other words, it allows engineering to function apart from the rest of the company.

Clark and Fujimoto (1991) found the following problems in conventional design: difficulty in designing for simplicity and reliability; excessive development times; weak design for producibility; inadequate attention to customers; weak links with suppliers; neglect of continuous improvement.

2.2.3 Discussion

Obviously, the transformation concept has been the foundation for product development and design management from the Second World War up to the 1980s. From the principles associated with the transformation concept, in particular the decomposition principle has been utilised. However, in a similar way to the situation in production, the transformation concept is not sufficient for the understanding or improvement of design processes. This is due to the bold idealisation inherent in this concept:

- There are also activities in design that do not contribute to transformation. For example, information is inspected, stored and communicated; these activities are not explicitly represented.
- Neither the total design process nor its parts are conceptually related to their customers.

Factually, the transformation view addresses only the first of the three questions posed by Turner earlier. In consequence the single-minded use of this view has contributed directly and indirectly to many persisting problems in design projects,[6] identified as anomalies earlier.

Clausing (1994) sees that the traditional design process has not moved far enough beyond partial design, that is, design from the point of view of one engineering discipline. Thus, according to Clausing (1994), the traditional approach suffers from failure of process (missing clarity with regard to the activities) and failure of co-operation (missing unity within the team). Solutions for these failures have been sought in the framework of CE, characterised (Rolstadås, 1995) as an endeavour for shortening lead time and for life cycle engineering.

In the following, it will be argued that it is no coincidence that various writers have recognised two motivations for or two fundamental aspects of CE.[7] This is because the principles and methods of CE are – predominantly implicitly[8] – based on two distinct concepts lacking from conventional approaches to design: the flow concept and the value generation concept (Figure 2.1).

2.3   Design as flow

2.3.1   Conceptualisation

If design is seen as a flow process, there are four different stages at which a piece of information may be: transformation, waiting, moving and inspection (Figure 2.1c). In fact, only transformation can be part of the design proper, other activities are basically not needed (and therefore called waste in industrial engineering), and should be eliminated rather than made more efficient. But a part of transformation, namely rework (or added work) due to errors, omissions, uncertainty, etc., is also waste.

In the literature on CE, this view[9] has been acknowledged, but not frequently. Augustin and Ruffer (1992) suggest using logistic thinking in the analysis of product development. Adachi *et al.* (1995) suggest conceiving CE as the application of JIT ideas to design. In their book on design improvement, Sekine and Arai (1994) focus on what happens to information in design: 'things are made through the flow of information'. The unit of

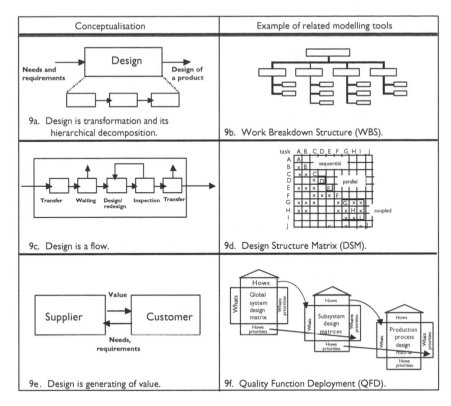

Conceptualisation	Example of related modelling tools
9a. Design is transformation and its hierarchical decomposition.	9b. Work Breakdown Structure (WBS).
9c. Design is a flow.	9d. Design Structure Matrix (DSM).
9e. Design is generating of value.	9f. Quality Function Deployment (QFD).

Figure 2.1 Transformation, flow and value generation view in design: conceptualisation and modelling.

Source: Koskela and Huovila (1997).

analysis is the total flow of information.[10] Reinertsen (1997) develops an approach on design management based on queuing theory.

In this view, improvement of design equals eliminating waste and related shortening of design time.[11] This view is significant because the amount of waste is large in any complex operation like engineering. When information flows are analysed in more detail, it is typically found that the share of transformation in the total flow time is very little. In general, the principles and methods of design waste elimination, to be analysed next, are related to the root cause of each waste category.

2.3.2 Rework

Cooper (1993) estimates that in complex electronic systems development projects, there are typically one to nine rework cycles. In design of large construction projects, there are typically from one-half to two and one-half rework cycles, according to Cooper. Reduction of this waste provides very worthwhile potential for improvement.

The major general cause for rework is variability associated with uncertainty (missing or unstable information). Thus, a variety of methods are required, in accordance with the nature of uncertainty. Especially, it is paramount to reduce aggressively uncertainty in the early phases of the engineering project (Bowen, 1992).

Changes in requirements or scope are disruptive for a product development and design project. Thus, it should be ensured that the scope is defined carefully, eliminating (avoidable) scope changes (Laufer, 1997).

Iterations may be needed due to constraints of downstream stages overlooked in upstream stages. This can be avoided by considering all life cycle phases simultaneously from the conceptual stage onwards. In practice, teamwork is often used for this purpose.

The need for iterations may also arise due to poor ordering of tasks. The Design Structure Matrix (DSM) method (Eppinger *et al.*, 1994) allows the representation of information flows between design tasks, and makes it possible to order the design tasks in such a way that the number of cases where a task has to send feedback to an earlier task is minimised (Figure 2.1d). Thus it is possible to minimise the waste due to unnecessary iterations. Also, a DSM analysis provides a starting point for scheduling, and helps to make the total design process transparent, which contributes to more effective design management.

Uncertainty may be due to intrinsic lack of definite information on matters under development. Prototyping, simulation etc. can be used to decrease this kind of technological uncertainty (Barkan *et al.*, 1992; Schrage, 1993). Uncertainty may also be reduced by decision. In later phases of the design project, especially, the design solution is often frozen in order not to complicate the realisation stage and its preparation.

Clearly, rework is also caused by the need for correcting design errors. Various tools of quality management can be used for reduction of errors.

2.3.3   *Transfer of information*

The time and effort needed for all the necessary transfer of information can be reduced through team approach, especially when the team is co-located (Reinertsen, 1997). In a team, much information can be transferred informally and orally, without paper or communication devices. Another option is in the elimination of vertical and horizontal divisions of labour and the resultant reduction in need for communication.[12] This means that the team is empowered to make decisions, which, earlier, were made by higher hierarchical layers.

2.3.4   Waiting for information

One reason for long waiting times of information is that output from each phase is transferred to the following phase in large batches (Reinertsen, 1997). Thus, splitting of design tasks, intense informal communication, and concurrence provide a solution to this. On the other hand, long waiting times may be due to poor control of the product design and development process, like too high a level of capacity utilisation (Reinertsen, 1997). Still another cause of waiting, especially in design for one-of-a-kind products, is the need to wait for customer decisions. This problem may be alleviated through better integration of customer decision making into the design process.

2.3.5   Unnecessary work

Design can also be conceived as pairs of supplier-customer. Poor specification of a supplier's work in relation to an internal customer's needs leads to added effort in the customer's activity, and also possibly to rework or continued work in the supplier's activity. Here, the consideration must be extended beyond design to manufacturing, which is the major internal customer of the design function. Several related methods (often called Design for X's) like Design for Manufacturability and Design for Assembly have emerged.

2.3.6   Technological solutions

Theoretically, the best solution is to eliminate a non-value-adding activity through new system structure, enhanced control or continuous improvement; for example, data transfer by collocation. If it is not possible to eliminate the non-value-adding activity, the 'second best' alternative is to make it more efficient. In this respect, various technological solutions for collaboration, engineering databases, etc. are instrumental, and, of course, increasingly important. On the other hand, information technology may provide new sources of waste. For example, non-compatibility of design tools causes one type of (set-up) waste: manual data conversion.[13]

2.3.7   Discussion

On the basis of the preceding considerations, it is justifiable to state that the majority of the prescriptions of CE can be explained through the flow concept. From the principles associated with this concept, those advocating waste elimination and variability reduction have been utilised.

2.4   Design as value generation

2.4.1   Conceptualisation

This view[14] focuses on value generated by a supplier to the customer(s). Value is generated through fulfilment of customer needs and requirements (Figure 2.1e). This fulfilment is carried out in a cycle, where customer requirements are captured and translated, through one or several stages, to a product or service delivered to the customer.

Product design comprises all stages where the functional features of a product are determined. In this cycle, at least three problems may emerge: requirements capture is not perfect; requirements get lost or remain unused; and translation is not optimal. The elimination of these problems,[15] to be discussed in the next sections, is the main focus of this view, and thus the source of improvement suggested.

Note that this view is analogous to mining, rather than manufacturing (as the previous view). The issue is to find the ore (requirements) and to have it processed so that no metal is rejected in slag (avoidance of value loss), and to produce an end result with as little impurity as possible (optimisation). However, often this mining metaphor is too simplistic because requirements do not necessarily exist at the outset but rather evolve in the product realisation process.

2.4.2.   Missing or evolving requirements

Why may part of the requirements be missed at the outset of the design? This may be due to a poor requirement analysis as such,[16] or specific features of the situation. One type of problem is due to the fact that the customer consists of a great number of people, and it is difficult to consolidate individual requirements into a coherent single set of requirements. Also, the number of requirements may be large or they may vary so much (Suh, 1995) that their management gets cumbersome. It has also been argued that problems in the early design stages may, sui generis, defy any attempt at predefinition (Green and Simister, 1996). In particular, regarding one-of-a-kind products, a certain evolution of requirements, reflecting changes in or enhanced understanding of customer needs, technology or manufacturing opportunities should be allowed (Ashton, 1992; Cusumano, 1997). Obviously, mass products, customised mass products and one-of-a-kind products present very different challenges to requirement capture.

The solution to this problem is a rigorous needs and requirements analysis at the outset in close co-operation with the customer(s). Several methods and tools have been developed for this purpose (Green and Simister, 1996; Reinertsen, 1997). For example, conjoint analysis helps in figuring out the customer priorities between requirements (Cook, 1997).

2.4.3 Loss of requirements

Another problem is that part of the requirements may be lost during the many-staged design process. For example, the design intent of a designer is not communicated for later steps, and may be spoiled by decisions in these (Fischer *et al.*, 1991). Requirements may be prioritised otherwise than meant by the customer.

For this problem, specific methods, like the Quality Function Deployment (QFD) method (Akao, 1990; Cohen, 1995), have recently emerged. It provides a formal linkage between requirements and corresponding solutions throughout the engineering (and production) process (Figure 2.1f). It also provides a systematic method of setting priorities, based on requirements as prioritised by the customer. Another, less formal tool is the method of Key Characteristics (Lee *et al.*, 1995). Key Characteristics attempt to identify and track features that significantly affect customer value. Thus, they provide a focus (rather than systematic elaboration, provided by QFD) on the most important product features.

2.4.4. Optimisation

Often one requirement has to be realised jointly by several product subsystems, designed by different specialists. Inversely, one subsystem has often to fulfil several requirements. Thus, optimisation in design consists of a myriad of trade-offs to be made wisely in the framework of global customer requirements. It is thus critical to know how relevant knowledge of individual designers can be enlarged (Yoshimura and Yoshikawa, 1998).

The method of QFD is instrumental also for optimisation. One important precondition for optimisation is teamwork together with such cultural features as commonly held goals, complete visibility, mutual consideration of all decisions, collaboration to resolve conflict and equality among discipline specialists (Linton *et al.*, 1992). The various methods of value engineering[17] or value analysis are also useful (Fowler, 1990).

The difficulty of catching all the variations in customer-use conditions, where the requirements should be fulfilled, was noted already by Shewhart[18] (1931). For creating products that consistently satisfy customer requirements, the Taguchi methods are instrumental (Taguchi, 1993; Clausing, 1994).

2.4.5 Discussion

Methods and tools instrumental from the point of view of the value generation concept have been developed both in the framework of the CE movement and in other professional communities. From the associated principles,[19] those stressing requirement capture and flow down as well as system capability have, in particular, been utilised in practice.

2.5   The TFV concept in design

2.5.1   Integration of the three concepts

A summary of all three views on design is provided in Table 2.1. It has to be noted that even if the three views have been presented as separate, they, in reality, exist as different aspects of design tasks. Each task in itself is a transformation. In addition, it is a stage in the total flow of design, where preceding tasks have an impact on it through timeliness, quality of output, etc., and it has an impact on subsequent tasks. Also, certain (external and internal) customer requirements direct the transformation of all input information into solutions in each task.

However, conventionally, it has only been the transformation view that has been explicitly modelled, managed and controlled. The other two views have been left for informal consideration by designers. The major contribution of CE is in extending modelling to the flow and value views, thus subjecting them to systematic management.

Table 2.1 Transformation, flow and value generation concepts of design

	Transformation concept	Flow concept	Value generation concept
Conceptualisation of design	As a transformation requirements and other input information into product design	As a flow of information, composed of transformation, inspection, moving and waiting	As a process where value for the customer is created through fulfilment of his requirements
Main principles	Hierarchical decomposition; control of decomposed activities	Elimination of waste (unnecessary activities); time reduction, rapid reduction of uncertainty	Elimination of value loss (gap between achieved value and best possible value), rigorous requirement analysis, systematised management of flow-down of requirements, optimisation
Methods and practices (examples)	Work Breakdown Structure, Critical Path Method, Organisational Responsibility Chart	Design Structure Matrix, team approach, tool integration, partnering	Quality Function Deployment, value engineering, Taguchi methods
Practical contribution	Taking care of what has to be done	Taking care that what is unnecessary is done as little as possible	Taking care that customer requirements are met in the best possible manner

How do the concepts interact? Do similar balancing issues arise as in production (Koskela, 2000)? Actually, our understanding of these questions is based on the predominance of the transformation view. Related empirical observations are discussed below.

First, it is a commonly occurring phenomenon that in task management, often too little time is reserved for needs analysis[20] and other issues of value management. This might be because value management is simply not conceptually captured in task management, based on the transformation view. However, poor definition of needs (domain of value management) causes disruption to task and flow management through untimely design changes.

Second, in traditional design, it is common practice for each task to produce a single design solution. In complex design situations, it is usual to iterate one alternative until a satisfactory solution emerges. It is assumed that each task can find the best solution in one shot. In fact, the transformation and flow views dominate in such practice, at the cost of the value view: the transformation view stresses getting each task done, and the flow view presupposes each activity to have a short and predictable duration. However, in the value view, the primary issue is in finding a still better solution for each task, using all the time available. This conventional practice, which can be called point-to-point design, is predominant in the current understanding of CE. However, recently, it has been pointed out that an alternative set-based type of CE is being used by Toyota (Ward *et al.*, 1995; Sobek and Ward, 1996). Here, designers explicitly communicate and think about sets of design alternatives. They gradually narrow the sets by eliminating inferior alternatives until a final solution emerges. Thus, set-based CE represents an approach in which the transformation, flow and value views are pursued in a more balanced way.

Third, in design task management, the need for a joint solution by designers of different disciplines, arising either from flow concept (i.e. a block of interrelated tasks in a DSM matrix) or value concept (different product subsystems contributing to one requirement) is usually not recognised (that is, there are no joint assignments) (Ballard and Koskela, 1998).

On the basis of these observations, it is justifiable to claim – like in the case of production – that in design management, the management needs arising from the three views should be integrated and balanced.

2.5.2   Implications

The lack of an adequate theory of engineering design is a major bottleneck, both for practice and research, including the information technology oriented endeavours (Fenves, 1996). Thus, further building, formalising and integration of the theory of design should be among the primary tasks of the design science community. We need a conceptual framework where all three approaches (transformation, flow and value) are integrated. This is needed especially in view of the pursuit of formal process models, used in computer-based description, analysis and simulation of engineering processes.

Let us take an example on the significance of theoretical understanding of design. In practice, establishing teams is often equated to CE. However, the results may be disappointing. Indeed, the teams in themselves are not a solution. More systematic flow and value generation processes would be the solution that, of course, is enabled by team organisation. Without ambitions and tools to model and manage the flow and value generation processes, team working degenerates into interaction for interaction's sake which does not correlate with performance, as Kahn (1996) has shown.

2.6   Conclusions

It is evident that CE is a practice in search of a theory. However, as shown earlier, the tools and methods of CE derive, implicitly or explicitly, from new conceptualisations of design, which thus provide the seed towards further development of the theory of design as well as new design management methods.[21] That the frontier of development is being called CE should not be allowed to hide the fact that the question is about design (and its management) in general rather than some particular type or aspect of design.

The historical development of design has many similarities to that of production. Originally, the first systematic attempts to manage design were based on the transformation concept, as in production. In the West, design anomalies caused by the idealisation error implied by the transformation concept were increasingly recognised in the 1980s – the same happened roughly simultaneously in production. Then CE emerged representing a similar theoretical shift to that in the case of lean production. The new methods of CE were based primarily on the flow concept but also on the value generation concept.

It can be thus argued that the TFV concept provides a theoretical foundation for design, too. Due to the intrinsic nature of design, the methods and practices are slightly different from those in production. The transformation view is instrumental in discovering which tasks are needed in an engineering undertaking. In the flow view, the basic thrust is to eliminate waste from the design processes. Thus, such practices as reduction of rework, team approach and releasing information for subsequent tasks in smaller batches are promoted. In the value generation view, the basic thrust is to reach the best possible value for the design solution from the point of the customer. Such practices as rigorous requirement analysis, systematised management of requirements and rapid iterations for improvement are put forward. These conceptualisations lead directly to the practices of CE, which can thus also be called theory-based design management.

2.7 Notes

1 Midler (1996) stresses the entrepreneurial nature of design in this era, as described by Schumpeter.

2 The presentation is based on earlier treatments of the topic by the author (Koskela and Huovila, 1997, 2000; Koskela, 2000).

3 It is interesting to note that rather similar factors stimulated systems engineering as concurrent engineering forty years later. According to Hall (1962), the emergence of systems engineering was due to growing complexity, expanding needs and environment, and shortage of manpower. Also the goals seem strikingly similar: 'systems engineering...attempts to shorten the time lags between scientific discoveries and their applications, and between the appearance of human needs and the production of new systems to satisfy these needs' (Hall, 1962).

4 Morris (1994, p. 217) comments that: '(W)hile the subject of "project management" is now comparatively mature, and recognized by thousands if not millions of managers as vitally important, it is in many respects still stuck in a 1960s time warp'.

5 In fact, the two definitions presented are slightly erroneous because in design, plenty of other input information is needed besides requirements.

6 Also, the CPM, when used in design and engineering management highlights the shortcomings of the transformation view. Because time has been abstracted away from the foundational concept of activity, it is difficult to present iterations in this method.

7 The characterisation by Rolstadås (1995) of CE as an endeavour for shortening of lead time and life cycle engineering matches these two failures to overcome (failure of process and failure of co-operation) well.

8 In contradiction to lean production, CE originally evolved solely through engineering practice (Sobek and Ward, 1996), rather than in interaction with new theoretical insights.

9 This perspective is adopted by Adler et al. (1994, 1996) who study the management of the development process in product development departments; thus their unit of analysis is not one project, as here, but rather the 'development project factory'.

10 Sekine and Arai (1994) argue that there are seven types of waste in design: preparing new drawings, retrieving or searching for drawings or material, permitting designers to set their own schedules, questioning unclear requirements and specifications, attending too many meetings and conferences, designing new estimate drawings and reference drawings and altering designs to correct defects.

11 Note that in design, shortening of lead time is much more an intrinsic goal than in production, where it is also, and often primarily, a means for cost reduction.

12 Soderberg (1989) notes: 'Like a manufacturing process with too many steps, an engineering organization with overly compartmentalized specialists builds up excess "WIP" between steps. The inevitable results: throughput delays and a rich supply of hidden problems that drive ineffective downstream activities.'

13 Surely, the solution put forward is standardisation of information structures but this has proven to be extremely difficult (Björk, 1995).

14 That this is a new perspective in design is supported by the following anecdote by Soderberg (1989): 'As an ex-chief engineer of a major new automobile model ruefully noted: "I discovered rather late that an engineer's design work is aimed at consumers so the final product can be marketed and purchased. For 20 years I thought engineers worked to create new designs."'

15 The solutions to these problems further clarify the implementation of principles of the value concept (Koskela, 2000), as seen from the point of view of product development and design. Especially, the principles on requirement capture, requirement flow-down and capability of the production system (here design subsystem of it) are involved.

16 Reinertsen (1997) comments: 'If there is one weakness in most product specification processes it is that the design team does not achieve an adequate understanding of the customer.'

17 Originally, value engineering focused on cost reduction. Although it is not uncommon to find this focus in current value engineering, modern value analysis looks rather at both the worth and cost of a product (Fowler, 1990).

18 Shewhart (1931, p. 356) says: 'Obviously, when equipment goes into the field it meets many and varied conditions, the influence of which on the quality of product is not in general known.'

19 An analysis of the whole product realisation process gave the following principles of value generation at hand (Koskela, 2000): ensure that all requirements get captured; ensure the flow down of customer requirements; take requirements for all deliverables into account; ensure the capability of the production system; measure value.

20 As indicated by Reinertsen (1997).

21 Examples of theory-based methods for design management are provided by Ballard (2002) and by Freire and Alarcon (2002).

2.8 References

Adachi, T., Enkawa, T. and Shih, L. C. (1995), 'A concurrent engineering methodology using analogies to just-in-time concepts', *International Journal of Production Research*, Vol. 33, No. 3, pp. 587–609.

Adler, P., Mandelbaum, A., Nguyen, V. and Schwerer, E. (1994), 'From project to process management in engineering: managerial and methodological challenges', in Dasu, S. and Eastman, C. (eds), *Management of Design*, Kluwer Academic Publishers, Norwell, pp. 61–82.

Adler, P. S., Mandelbaum, A., Nguyen, V. and Schwerer, E. (1996), 'Getting the most out of your product development process', *Harvard Business Review*, March–April, pp. 134–152.

Akao, Yoji (ed.) (1990), *Quality Function Deployment*, Productivity Press, Cambridge, MA, 369 p.

Ashton, J. A. (1992), 'Managing design for continuous improvement in a system job shop', *Manufacturing Review*, Vol. 5, No. 3, pp. 149–157.

Augustin, S. and Ruffer, J. (1992), 'How to reduce lead-time in R & D: the application of logistic thinking on the "one-of-a-kind" -business', in Hirsch, B. E. and Thoben, K.-D. (eds), *'One-of-a-kind Production': New Approaches*, Elsevier Science, Amsterdam, pp. 255–262.

Ballard, G. (2002), 'Managing work flow on design projects: a case study', *Engineering Construction and Architectural Management*, Vol. 9, No. 3, pp. 284–291.

Ballard, G. and Koskela, L. (1998), 'On the agenda of design management research', *6th Annual Conference of the International Group for Lean Construction*, Guaruja, Sao Paulo, Brazil, 13–15 August (available at: http://www.ce.berkeley.edu/~tommelein/IGLC-6/index.html, January 2005).

Barkan, P., Iansiti, M. and Clark, K. (1992), *Prototyping as a Core Development Process*, Draft. Mimeo. 17 p.

Björk, B.-C. (1995), *Requirements and Information Structures for Building Product Data Models*, VTT Publications 245, Technical Research Centre of Finland, Espoo. 88 p. + app. 75 p.

Bowen, H. (1992), 'Implementation projects: decisions and expenditures', in Heim, J. and Compton, W. (eds), *Manufacturing Systems: Foundations of World-Class Practice*, National Academy Press, Washington, DC, pp. 93–99.

Carter, D. and Baker, B. (1992), *Concurrent Engineering: The Product Development Environment for the 1990s*, Addison-Wesley, Reading, PA.

Clark, K. and Fujimoto, T. (1991), *Product Development Performance*, Harvard Business School Press, Boston, MA, 409 p.

Clausing, D. (1994), *Total Quality Development*, ASME Press, New York, 506 p.

Cohen, L. (1995), *Quality Function Deployment*, Addison-Wesley, Reading, 348 p.

Cook, H. E. (1997), *Product Management – Value, Quality, Cost, Price, Profit and Organization*, Chapman & Hall, London, 411 p.

Cooper, K. (1993), 'The rework cycle: benchmarks for the project manager', *Project Management Journal*, Vol. XXIV, March, pp. 17–21.

Cusumano, M. (1997), 'How Microsoft makes large teams work like small teams', *Sloan Management Review*, Fall, pp. 9–20.

Dasu, S. and Eastman, C. (eds) (1994), *Management of Design*, Kluwer Academic Publishers, Norwell, MA, 277 p.

Eppinger, S. D., Whitney, D. E., Smith R. P. and Gebala, D. A. (1994), 'A model-based method for organizing tasks in product development', *Research in Engineering Design*, Vol. 6, pp. 1–13.

Fenves, S. J. (1996), 'The penetration of information technologies into civil and structural engineering design: state-of-the-art and directions toward the future', in Kumar, B. and Retik, A. (eds), *Information Representation and Delivery in Civil and Structural Engineering Design*, Civil Comp Press, Edinburgh, pp. 1–5.

Fischer, G., Lemke, A. C., McCall, R. and Morch, A. I. (1991), 'Making argumentation serve design', *Human-Computer Interaction*, Vol. 6, No. 3–4, pp. 393–419.

Fowler, T. (1990), *Value Analysis in Design*, Van Nostrand Reinhold, New York, 302 p.

Freire, J. and Alarcon, L. F. (2002), 'Achieving lean design process: improvement methodology', *Journal of Construction Engineering and Management*, May/June, pp. 248–256.

Green, S. D. and Simister, S. J. (1996), 'Group decision support for value management', in Langford, D. A. and Retik, A. (eds), *The Organization and Management of Construction. Volume 2*, E & FN Spon, London, pp. 529–540.

Hall, A. D. (1962), *A Methodology for Systems Engineering*, Van Nostrand, Princeton, NJ, 478 p.

Hubka, V. and Eder, W. E. (1988), *Theory of Technical Systems*, Springer, Berlin, 275 p.

Jo, H. H., Parsaei, H. R. and Sullivan, W. R. (1993), 'Principles of concurrent engineering', in Parsaei, H. R. and Sullivan, W. G. (eds), *Concurrent Engineering*, Chapman & Hall, London, pp. 3–23.

Kahn, K. B. (1996), 'Interdepartmental integration: a definition with implications for product development performance', *Journal of Product Innovation Management*, Vol. 13, No. 2, pp. 137–151.

Koskela, L. (1992), *Application of the New Production Philosophy to Construction*, Technical Report; 72, Center for Integrated Facilities Engineering (CIFE), Stanford University, 75 p.

Koskela, L. (2000), *An Exploration Towards a Production Theory and its Application to Construction*, VTT Publications; 408, VTT Building Technology, Espoo, 296 p. (available at http://www.inf.vtt.fi/pdf/publications/2000/P408.pdf, January 2005).

Koskela, L. and Howell, G. (2002), 'The underlying theory of project management is obsolete', in Slevin, D. P., Cleland, D. I. and Pinto, J. K. (eds), *Proceedings of PMI Research Conference*, Project Management Institute, pp. 293–301.

Koskela, L. and Huovila, P. (1997), 'Foundations of concurrent engineering', in Anumba, C. J. and Evbuomwan, N. F. O. (eds), *Concurrent Engineering in Construction – Papers presented at the 1st International Conference*, London, 3–4 July 1997, pp. 22–32.

Koskela L. and Huovila, P. (2000), 'On foundations of concurrent engineering', *International Journal of Computer Integrated Design and Construction*, Vol. 2, No. 1, pp 2–8.

Laufer, A. (1997), *Simultaneous Management: Managing Projects in a Dynamic Environment*, Amacom, New York, 313 p.

Lee, D. J., Thornton, A. C. and Cunningham, T. W. (1995), 'Key characteristics for agile product development and manufacturing', in Barker, J. J. (ed.), *Creating the Agile Organization: Models, Metrics and Pilots*, Agility Forum Conference Proceedings, Vol. 2, pp. 258–268.

Linton, L., Hall, D., Hutchinson, K., Hoffman, D., Evanczuk, S. and Sullivan, P. (1992), *First Principles of Concurrent Engineering*, CALS/Concurrent Engineering Working Group – Electronic Systems, Litton Amecon, College Park, 182 p.

Methodik zum Entwickeln und Konstruieren technischer Systeme und Produkte. (1986), Richtlinie VDI 2221. VDI-Verlag, Düsseldorf, 35 p.

Midler, C. (1996), Modèles gestionnaires et régulations économiques de la conception. Cahiers du Centre de Recherche en Gestion No. 13, Du Management de Projet aux Nouvelles Rationalisations de la Conception, CNRS, Ecole Polytechnique, pp. 11–26.

Mistree, F., Smith, W. and Bras, B. (1993), 'A decision-based approach to concurrent design', in Parsaei and Sullivan (eds), *Concurrent Engineering: Contemporary Issues and Modern Design Tools*, Chapman & Hall, pp. 127–158.

Morris, P. (1994), *The Management of Projects*, Thomas Telford, London, 358 p.

Murmann, P. (1994), 'Expected development time reductions in the German mechanical engineering industry', *Journal of Product Innovation Management*, Vol. 11, pp. 236–252.

Prasad, B. (1996), *Concurrent Engineering Fundamentals*, Vol. 1, Prentice Hall, Upper Saddle River, NJ, 478 p.

Project Management Institute (1996), *A Guide to the Project Management Body of Knowledge*, 176 p.

Putnam, A. O. (1985), 'A redesign for engineering', *Harvard Business Review*, May–June, pp. 139–144.

Reinertsen, D. G. (1997), *Managing the Design Factory*, The Free Press, New York, 267 p.

Rolstadås, A. (1995), 'Planning and control of concurrent engineering projects', *International Journal of Production Economics*, Vol. 38, pp. 3–13.

Schrage, D. (1993), 'Concurrent design: a case study', in Kusiak, A. (ed.), *Concurrent Engineering: Automation, Tools, and Techniques*, John Wiley & Sons, pp. 535–580.

Sekine, K. and Arai, K. (1994), *Design Team Revolution: How to Cut Lead Times in Half and Double Your Productivity*, Productivity Press, 305 p.

Shewhart, W. A. (1931), *Economic Control of Quality of Manufactured Product*, Van Nostrand, New York, 501 p.

Sobek, D. K. II. and Ward, A. C. (1996), 'Principles from Toyota's set-based concurrent engineering process', *Proceedings of the 1996 ASME Design Engineering Technical Conferences and Computers in Engineering Conference*, 18–22 August, Irvine, California, 9 p.

Soderberg, L. G. (1989), 'Facing up to the engineering gap', *The McKinsey Quarterly*, Spring, pp. 2–18.

Suh, N. P. (1995), 'Design and operations of large systems', *Journal of Manufacturing Systems*, Vol. 14, No. 3, pp. 203–213.

Taguchi, G. (1993), *Taguchi on Robust Technology Development*, ASME Press, New York, 136 p.

Tomiyama, T. (1995), 'A Japanese view on concurrent engineering', *Artificial Intelligence for Engineering Design, Analysis and Manufacturing*, Vol. 9, pp. 69–71.

Turner, J. R. (1993), *The Handbook of Project-based Management*, McGraw-Hill, London, 540 p.

Ward, A., Liker, J. K., Cristiano, J. J. and Sobek, D. K. II. (1995), 'The second Toyota paradox: how delaying decisions can make better cars faster', *Sloan Management Review*, Spring, pp. 43–61.

Yoshimura, M. and Yoshikawa, K. (1998), 'Synergy effects of sharing knowledge during cooperative product design', *Concurrent Engineering: Research and Applications*, Vol. 6, No. 1, pp. 7–14.

Chapter 3

Readiness assessment for Concurrent Engineering in construction

Malik M. A. Khalfan, Chimay J. Anumba and Patricia M. Carrillo

3.1 Introduction

The UK Government initiated reports such as the Latham Report (1994) and the Egan Report (1998) have recommended the improvement of the construction industry's business performance. The need for greater co-ordination and integration within the industry has led to the adoption of various concepts from other industries. One of these, which offers major scope for effective co-ordination and integration within the industry, is Concurrent Engineering (CE) (Kamara *et al.*, 2000). CE sometimes called simultaneous engineering or parallel engineering, has been defined in several ways by different authors. The most popular definition is that by Winner *et al.* (1988), who state that CE '… is a systematic approach to the integrated, concurrent design of products and their related processes, including manufacture and support. This approach is intended to cause the developers, from the outset, to consider all elements of the product life cycle from conception through disposal, including quality, cost, schedule, and user requirements'. In the context of the construction industry, Evbuomwan and Anumba (1998) define CE as an 'attempt to optimise the design of the project and its construction process to achieve reduced lead times, and improved quality and cost by the integration of design, fabrication, construction and erection activities and by maximising concurrency and collaboration in working practices'. This is in sharp contrast with the traditional approach to construction project delivery.

In order to introduce aspects of CE in the construction project delivery process, various research efforts have been undertaken. These include *ToCEE*, which focused on developing information exchange systems that support a CE environment over the building lifecycle (ToCEE, 1997); *CICC*, which was concerned with enabling communication across the whole of construction project and at all stages of the lifecycle (Duke and Anumba, 1997); *CONCUR*, which focuses on electronic information exchange from the inception to tendering and construction planning stage (CONCUR, 1999); *COMMIT*, which addresses the issues of integration and collaboration by efficient information management (Rezgui et al., 1997); *DESCRIBE*, which focuses on the development of software to facilitate concurrent storage, access, and modification of design information, irrespective of the location of the designer (Carnduff et al., 1997); and *IDS*, which deals with the integration of various tools for the concurrent design and fabrication of steel structures (Wailes et al., 1997). A detailed account of these efforts is compiled and presented by Kamara et al. (2000). They have concluded that much more needs to be done if the reported benefits of CE in other industries such as manufacturing can be realised in construction industry. It is also concluded that an important aspect of CE implementation in the construction industry, which is often overlooked, is the need to carry out a readiness assessment of the construction supply-chain for CE implementation. This is expected to establish the level of CE maturity of different sectors of the supply-chain with a view to informing implementation efforts. Therefore, in order to establish the level of maturity and improve planning for CE implementation, the construction industry needs a specific readiness assessment model (Khalfan and Anumba, 2000a; Khalfan et al., 2001).

This chapter compares the existing CE readiness assessment tools and models, examines their appropriateness for the construction industry in the light of current practices within the industry, and discusses the development and implementation of a new readiness assessment model (the BEACON model) for the construction industry.

3.2   CE readiness assessment

3.2.1   Introduction

As discussed in the previous section one approach that has been successfully used to improve CE implementation planning is to conduct a readiness assessment of an organisation prior to the introduction of CE. This helps to investigate the extent to which the organisation is ready to adopt CE (Componation and Byrd, 1996), and to identify the critical risks involved in its implementation within the company and its supply chain. CE readiness assessment has been successfully used for the planning of CE implementation in several industry sectors, notably manufacturing and software engineering.

3.2.2   Comparison of readiness assessment tools and models

There are several tools and models, which are being used for readiness assessment of organisations for CE (CERC Report, 1993; de Graaf and Sol., 1994; Bergman and Ohlund, 1995; Wognum et al., 1996; Kwak and Ibbs, 1997; Aouad et al., 1998; SPICE Questionnaire, 1998;

Brookes *et al.*, 2000; Finnemore and Sarshar, 2000). A comparison of these models and tools is presented in Table 3.1.

3.2.3   Framework for comparison

The framework for comparison discusses the characteristics of the available tools and models under a number of generic criteria, which include:

- Aspects covered: this highlights the main issues addressed in each tool.
- The status of the tool/method: this shows the current standing of the tool/model in terms of whether it is a research prototype, commercial tool or currently under development.
- Survey method: this identifies how the data collection is carried out – that is either by questionnaires, interviews or both.
- Software availability: this identifies those tools and models which are accompanied by a software that can be used during the readiness assessment.
- Ease of use: an indication of the user-friendliness of the tools/models.
- Can be used for concurrent engineering readiness assessment: this identifies the tools and models which can be used for CE readiness assessment.
- Appropriateness for use in construction: this identifies the tool/model suitability for the construction industry.

3.2.4   Findings

From the comparative analysis (Table 3.1), it could be concluded that most of the tools and models address improvements in the product development process, and the use of technology to facilitate the development process. Some of the tools and models also cover the organisational environment to support the development process. The status of the tools and models shows that most of them are under development with only very few being used on a commercial basis. With regard to software availability, there are only a few tools and models which are accompanied by their own software. Many of the tools and models are easy to use and user-friendly. Most of the tools and models reviewed were developed to assess the product development process within an organisation. However, they can also be used as a CE readiness assessment tool with appropriate modification. A few of them were designed for CE readiness assessment. An assessment of the use of these tools and models within the construction industry shows that none of the tools and models is ideally suited for use in construction (Khalfan and Anumba, 2000a).

Table 3.1 Comparison of CE readiness assessment and related models

Tools/Models→ Criteria↓	RACE	PMO	PMO-RACE	PRODEVO	CMM	SPICE	(PM)^2	SIMPLOFI
Aspects covered	Process • Customer focus • Product assurance • Leadership • Team formation • Strategy deployment • Agility • Teams within the organisation • Process focus • Management system • Discipline Technology • Project architecture • Application tools • Communication • Co-ordination • Information	Organisational environment • Task environment • General environment Processes • Primary processes • Control processes: Strategic level, Adaptive level, and Operational level • Support processes	Aspects covered are the same as PMO and RACE because this is the combination of both of these tools	• Customer and user focus • Process focus • Team and project focus • Life-cycle perspective • Communication	Process • Pre-project phase • Pre-construction phase • Construction phase • Post-construction phase Information technology • Simulation • Integration • Intelligence • Communications • Visualisation • IT support	• Brief management • Project planning • Project tracking and monitoring • Contract management • Quality assurance • Project change management • Risk management • Organisation process focus • Organisation process definition • Training programme • Inter-disciplinary Co-ordination	• Planning to execute a project • Definition of project activities • Cost estimates for the project • Project Management (PM) process • PM-related data collection and analysis • Utilisation of PM tools and techniques • Working as a team • Senior management support	• The structure of teams • Control mechanisms (whether control mechanisms should reside with functions or projects) • The degree to which the process should be parallelised • How specialised people operating the process should be • The degree of automation in the tools used

(Table 3.1 continued)

Table 3.1 Continued

Tools/ Models→ Criteria↓	RACE	PMO	PMO-RACE	PRODEVO	CMM	SPICE	(PM)²	SIMPLOFI
	sharing • Integration					• Peer review • Technology management		
Status of tool/ method	Commercial	Development ongoing	Development ongoing	Development ongoing	Commercial	Research prototype	Development ongoing	Commercial
Survey method	Questionnaire and interview	Interviews and description of current projects, formal procedures and quality hand book	Questionnaire and interview	Questionnaire	Questionnaire and interview	Questionnaire and semi-structured interview	Questionnaire	Questionnaire
Software availability	Yes, also uses other software (e.g. SPSS)	Can use any modelling software	Yes	None	Yes, but also use other software (e.g. SPSS)	None	None	Yes
Ease of use	Yes, but technological aspect is complicated to answer and is only for specialists	Yes, but seemed to be incomplete, that's why merged with RACE later on	Yes, and it seems to be completed after the combination of PMO and RACE	Yes	Yes	Yes, MCQs are developed with additional space for comments	Yes	Yes, user-friendly software

	Yes, basically made for this purpose	Basically used for analysing and designing organisations	Yes, mainly for readiness assessment but also used for CE implementation process	Basically developed for assessing CE process	Yes, but basically used for CE implementation process	Basically used for process improvement	Basically used as a yardstick for an organisation applying PM practices and processes	Basically used to assist those, who are responsible for product introduction within an organisation
Can be used for CE readiness assessment?	Yes, basically made for this purpose	Basically used for analysing and designing organisations	Yes, mainly for readiness assessment but also used for CE implementation process	Basically developed for assessing CE process	Yes, but basically used for CE implementation process	Basically used for process improvement	Basically used as a yardstick for an organisation applying PM practices and processes	Basically used to assist those, who are responsible for product introduction within an organisation
Appropriateness for use in construction	Yes, but requires some modifications	Yes, but basically used for analyzing and designing an organisation, its process and technology	Yes, but RACE model requires modification before applying to construction	Yes, but it requires changes to address construction specifically	Yes, but basically developed for software industry, therefore it requires changes before applying to construction	Yes, but this tool is basically made for process improvement within construction projects	Yes, but this tool is basically developed to determine and to position an organisation's relative PM level with other organisations	Yes, but this tool focuses on the introduction of one specific product in an organisation. Therefore, in any construction organisation, it can be used for a specific project and it would give the position of the project and not the position of the organisation.

Source: Khalfan and Anumba (2000c).

3.2.5   The move towards CE in construction

As already mentioned earlier that CE is an attempt to optimise the process of project design and construction in order to achieve shorter lead times, and improved quality and cost. This is actually achieved through the integration of design, and construction activities and by increasing simultaneous activities and collaboration among the construction supply chain participants. This is in sharp contrast with the traditional approach to construction project delivery.

3.2.6   Traditional approach

In the construction industry, based on the client brief, the architect produces an architectural design, which is given to the structural engineer, who on completing the structural design passes the project to the quantity surveyor to produce the costing and bill of quantities. This goes on until the project is then passed on to the contractor who takes responsibility for the construction of the facility. This scenario, which is similar to the 'over the wall' approach (Evbuomwan and Prasad, 1997; Anumba, 1998), is shown in Figure 3.1. The key disadvantages prevalent with this approach include:

- The fragmentation of the different participants in the construction project, leading to misperceptions and misunderstandings;
- The fragmentation of design and construction data, leading to design clashes, omissions and errors;
- The occurrence of costly design changes and unnecessary liability claims, occurring as a result of the above;
- The lack of true life-cycle analysis of the project, leading to an inability to maintain a competitive edge in a changing marketplace;

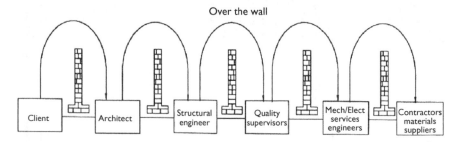

The traditional design and construction process

Figure 3.1 The 'over the wall' approach.

Source: Evbuomwan and Anumba (1998).

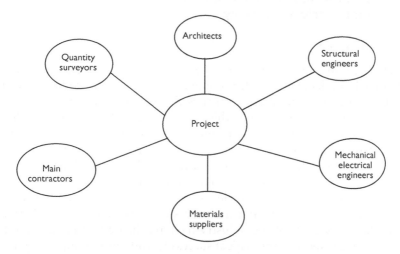

Figure 3.2 An integrated project team.

Source: Evbuomwan and Anumba (1998).

- Lack of communication of design rationale and intent, leading to design confusion and wasted effort.

To address these issues, there is an urgent need for a paradigm shift within the construction industry. This should involve the adoption of new business strategies, with the aim of integrating the functional disciplines (see Figure 3.2) at the early stages of the construction project (Evbuomwan and Anumba, 1998).

3.2.7   Application of CE to construction

There is a need to improve the performance of the construction supply chain. This can be achieved especially during the design process by considering all aspects of the project's downstream phases concurrently. This would result in incorporation of requirements from the construction, operation and maintenance phases at an early stage of a project and would undoubtedly lead to an overall improvement in project performance. The essential constituents of 'Concurrent Construction' are as follows (Love and Gunasekaran, 1997):

- The identification of associated downstream aspects of design and construction processes.
- The reduction or elimination of non-value-adding activities.
- The development and empowerment of multi-disciplinary teams.

3.3   CE readiness assessment of the construction industry

3.3.1   The need

As discussed in previous sections, CE Readiness Assessment is used to improve CE implementation. It is conducted before the introduction of CE within an organisation, and investigates the extent to which the organisation is ready to adopt CE. While this has been carried out in other industry sectors, it is unusual for such assessments to be undertaken in construction supply chains. Furthermore, Muya *et al.* (1999) show that current industry practices do not support integration of the whole supply chain during the construction process. It is therefore imperative that, for CE implementation in the construction industry to deliver the expected benefits, a readiness assessment of the construction industry should be undertaken. This will ensure that all sectors of the industry have reached an acceptable level of maturity with respect to the critical success factors for CE implementation, and may indicate the likelihood of the following benefits:

- Better and more effective CE implementation within the construction industry;
- Enabling the industry to evaluate and benchmark its project delivery processes;
- Development of more appropriate tools for CE implementation within the industry;
- Enabling the industry to identify areas which require improvements or changes;
- Enabling the industry to realise the need for CE implementation in order to bring about improvements in the whole project delivery process.

3.3.2   Choice of an assessment model for construction

After analysing the comparison matrix (see Table 3.1), RACE would appear to be the most appropriate for use as the Readiness Assessment Tool for Concurrent Engineering in the construction industry for the following reasons:

- Aspects covered in the RACE model such as customer focus, team formation, management systems, communication and integration systems, etc., can be used for CE readiness assessment in the construction industry with some modification.
- Commercial usage of the RACE model makes it more reliable.
- The RACE model questionnaire addresses and assesses similar critical business drivers to those used in the construction industry.

- RACE is essentially a CE readiness assessment model, thus it is more appropriate than other tools and models, which were developed to assess the project/product development process within an organisation.

However, the RACE model requires adaptation and modification for use in the construction industry. This is because RACE was developed for readiness assessment for CE in other industries such as manufacturing and software engineering industry. Thus, it needs to be tailored to the specific requirements of the construction industry and the people working within the industry. The following are some of the reasons which indicate that RACE in its current form is not suitable for the construction domain and therefore, requires modification:

- RACE was designed for assessing the readiness of other industries such as software, automotive, manufacturing and electronic industries, all of which have different characteristics to construction.
- Aspects covered focus on the processes in the above mentioned industries and require changes to assess the construction process.
- The structure of teams within the above mentioned industries are different from typical construction project teams.
- The level of technology usage in the afore-mentioned industries is different from that in the construction industry.
- The products of the other industry sectors satisfy a large number of customers whereas a construction project is one-off in nature, typically fulfilling the needs of a particular client or organisation.
- The level of integration, communication, co-ordination and information sharing are different between construction and the above-mentioned industries.
- Managing a manufacturing product and a construction project require different levels of management skills.

3.4   Development of a model for construction

3.4.1   Background

A CE readiness assessment model has been developed by the authors for assessing the construction industry. The initial version of the BEACON model, which is shown in Figure 3.3, was developed with an associated questionnaire from the RACE model. The proposed model had similarities with the RACE in terms of the key assessment elements (i.e. most of them cover the same issues), questionnaire criteria and diagrammatic representation (spider or radar diagram). However, it differs from the RACE model in that it focuses specifically on construction processes. The model was initially divided into two sections or aspects (as shown in Figure 3.1), the

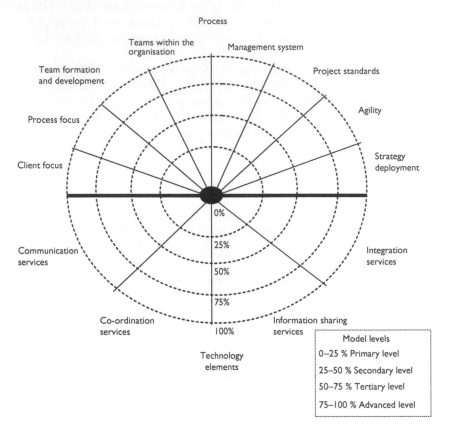

Figure 3.3 CERAM construct model.
Source: Khalfan and Anumba (2000a).

upper half presents eight process-related elements and the lower half contains four technology-related issues. The process aspect includes the client focus regarding the project, improvement in the construction process itself, formation and development of teams for carrying out project tasks, improving the management systems of the organisation, maintaining the project and process standards, bringing agility into the construction process, and employing and exploiting project strategy. The technology aspect includes the services related to communication, co-ordination, information sharing and integration (Khalfan and Anumba, 2000b).

Development of the model was carried out in several steps. A literature review of CE in other industries was carried out which identified the critical success factors and pit-falls during CE implementation. The next step was the review of CE readiness assessment models used in other industries; this included a comparative study, which is summarised in Table 3.1. These steps then led to the development of the BEACON model and its associated questionnaire. Before using the model for the assessment within the construction industry, a pilot study was carried out for both the model and its associated questionnaire. The purpose of the pilot study was to validate the model and its associated questionnaire, and obtain feedback for further refinement of the model and its associated questionnaire. The pilot study was carried out with three construction organisations, whose senior management staff filled in the assessment questionnaire. The results of the pilot study suggested areas for improvement within the questionnaire and the model itself. The pilot study also revealed the following limitations of the model:

- Inadequate focus on people and product in the model.
- The four-level assessment scale of the model meant that there was no neutral or middle level.
- Coarse-grained model division in terms of number of elements assessed.

Therefore, in order to incorporate the feedback from the pilot study and overcome the limitations, the model was refined and modified, resulting in its refined form, presented in Figure 3.4.

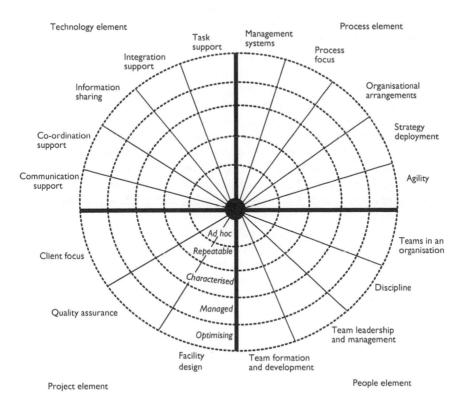

Figure 3.4 BEACON model.

3.4.2   The BEACON model

The BEACON model (see Figure 3.4) is divided into four quadrants or sections to represent four elements or aspects of the model, which are Process, People, Project and Technology. The first quadrant contains five critical process factors used to assess the process maturity level of a construction organisation. The second quadrant contains four critical people factors used to assess the team level issues within the organisation while the third quadrant is comprised of three critical project factors used to assess the client's requirement and design related issues. The fourth quadrant presents five technology related critical factors used to characterise the introduction and utilisation of advanced tools and technology within the organisation. The key advantage of the model is that it does not only include the process and the technology aspects as covered in other models but also introduces two new dimensions, people and project elements. These elements were covered to a limited extent in existing readiness assessment models and tools but were not adequately emphasised. The rationale behind including the people and the project elements is that both of them are as critical as the process and the technology elements and should be distinguished Martin and Evans, 1992; Crow, 1994; Chen, 1996; Brooks and Foster, 1997; Love and Gunasekaran, 1997; Paul and Burns, 1997; Ainscough and Yazdani, 1999; Al-Ashaab and Molina, 1999; Young, 1999; Khalfan and Anumba, 2000a). This is one of the novel features in the BEACON model.

For all of the elements, five levels have been adopted from the RACE model (CERC Technical Report, 1992), which indicate the level of maturity of an organisation with respect to the quality of project development process, team-working, completed project itself and technology employed within the organisation. These five levels are Ad hoc, Repeatable, Characterised, Managed and Optimising and are described in Table 3.2. The Ad hoc level indicates that an organisation is not aware of CE practices or is not ready to adopt CE whereas Optimising level shows that the organisation is ready to adopt CE or is already practising CE within its project delivery process.

A model-based questionnaire (called the BEACON Questionnaire) has been developed for use in assessing construction organisations such that the elements covered in this model would be assessed using this questionnaire. The assessment scale has five possible options: 'Always', 'Most of the Time', 'Sometimes', 'Rarely' and 'Never'. The BEACON Questionnaire can be used for assessing CE readiness of:

(a) A static construction organisation, for example an architectural or construction organisation, etc., which has different teams for different on-going projects, and

(b) A virtual construction organisation, which consists of various members from different construction organisations, forming a Project Development Team (PDT) and working on a single project (Khalfan, 2000). Figure 3.5 illustrates the PDT and its sub-teams, which may be responsible for supervising the whole project development process from inception until hand-over.

Table 3.2 BEACON model maturity levels (adopted from RACE model)

Maturity level	Description
Ad hoc	This level is characterised by ill-defined procedures and controls, and by confused and disordered teams that do not understand their assignment nor how to operate effectively. Informal interaction with the client is observed, management of the project development process is not applied consistently in projects and modern tools and technology are not used consistently
Repeatable	Standard methods and practices are used for monitoring the project development process, requirements changes, cost estimation, etc. The process is repeatable. There are barriers to communicate within the project development team. Interaction with the client is structured but it is only at the inception of the project. Minimal use of computer and computer-based tools
Characterised	The project development process is well characterised and reasonably well understood. A series of organisational and the process improvements have been implemented. Teams may struggle and fall apart as conflicts are addressed but a team begins to respect individual differences. Most individuals are well aware of client's requirements but client is not involved in the process. Moderate use of proven technology for increasing group effectiveness
Managed	The project development process is not only characterised and understood but is also quantified, measured and reasonably well controlled. Tools are used to control and manage the process. The uncertainty concerning the process outcome is reduced. Work is accomplished by the project development team and conflicts are addressed. Client is involved throughout the process. Appropriate utilisation of available technology and computer-based tools
Optimising	A high degree of control is used over the project development process and there is a major focus on significantly and continually improving development operations. Team performance is regularly measured, and performance measures are continuously validated. Client is a part of project development team from inception and all project decisions are prioritised based on client's needs. Optimal utilisation of appropriate plant and technology and technology-mediated group work is observed

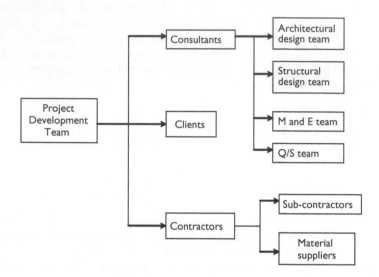

Figure 3.5 Typical team structure within a virtual construction organisation.
Source: Khalfan and Anumba (2000b).

3.4.3   CE readiness assessment case studies

Case studies were carried out by using the BEACON model in order to assess the CE readiness of the UK construction industry. One of the reasons for carrying out case studies is the fact that they help to solve current problems through an examination of what has happened in the past and what is happening now, and thus save a lot of time (Yin, 1989). For the purpose of the case studies, the industry was divided into five categories: clients, consultants, contractors, sub-contractors and material suppliers.

3.4.4   Methodology

Ten companies within each category were selected randomly with the expectation that at least five of them would respond. Questionnaires were sent out with a covering letter to all the selected companies. Before sending out the questionnaires, each company was contacted and the most appropriate person was identified, either from senior or middle management, who had knowledge about the company and could adequately complete the questionnaire. A summary of the assessment results is complied and presented in Table 3.3, which shows average percentages for all the elements within each category. The average percentages for each factor within the elements were calculated after assessing the questionnaire responses for

Table 3.3 Summary of the readiness assessment results

Elements	Construction supply chain participants				
	Clients (%)	Consultants (%)	Contractors (%)	Sub-contractors	Material suppliers and manu-facturers (%)
Process element (Avg)	*68.13*	*71.69*	*73.94*	*80.04*	*63.15*
Management systems	66.13	71.64	77.31	83.33	54.98
Process focus	70.36	66.83	70.38	82.69	63.26
Organisational framework	68.33	68.75	78.00	81.67	58.75
Strategy deployment	73.75	77.50	74.50	76.67	67.50
Agility	62.08	73.75	69.50	75.83	71.25
People element (Avg)	*68.56*	*75.39*	*78.81*	*81.13*	*71.88*
Team formation and development	70.42	71.88	76.50	86.67	81.25
Team leadership and management	81.25	75.78	81.88	84.38	67.71
Discipline	66.67	80.47	85.63	87.50	80.21
Teams In an organisation	55.91	73.44	71.25	65.97	58.33
Project element (Avg)	*76.92*	*73.59*	*76.60*	*85.51*	*73.08*
Client focus	80.89	65.91	69.09	82.58	72.73
Quality assurance	69.79	81.26	86.26	90.63	72.92
Facility design	80.09	73.61	74.44	83.33	73.61
Technology element (Avg)	*55.01*	*52.81*	*67.56*	*76.11*	*42.32*
Communication support	57.92	60.63	64.50	83.33	55.83
Co-ordination support	49.30	39.58	62.78	72.22	35.18
Information sharing	55.69	50.00	70.00	65.15	44.04
Integration support	55.76	48.44	69.38	82.30	40.63
Task support	56.40	65.39	71.15	77.56	35.90

each category. A brief account of all case studies within each category is presented in the following sub-sections, with the results plotted on the BEACON model diagram for each industry sector.

3.4.5   Readiness of clients

Thirty-three per cent of client organisations responded to the questionnaire, ranging from large to small in size and representing different client sectors such as hospitals, academic institutions, etc. All respondents identified the people element as the most important and the technology element as the least important element from their point of view. The average assessment result is plotted on the BEACON model diagram shown in Figure 3.6. The clients are doing best in the project element, need the most improvements in the technology element, and have average performance under the process and people elements. The overall result of client organisations shows that some of the critical factors are at the 'managed level' while the rest are at the 'characterised level' of CE readiness. This confirms that the client organisations are not ready to adopt CE and the areas which need attention are: all factors within the technology element, agility within the process element, teams in an organisation within the people element and quality assurance within the project element.

3.4.6   Readiness of consultants

The response rate for consultants was the same as for clients, that is four consulting organisations out of twelve architecture and engineering consultants responded to the questionnaire. Most of the respondents stated that the people element is the most important and the technology element the

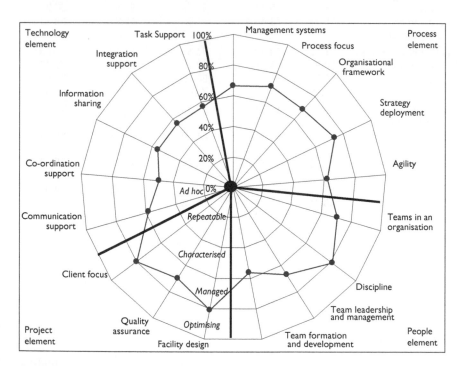

Figure 3.6 CE readiness of clients.

least important element for them. The average readiness assessment result for consultants is shown in Figure 3.7. This shows that consulting organisations are at the 'managed level' except for some of the critical factors, which indicate the 'characterised level' of the CE readiness for the organisations. Most of the critical factors in the process, people and project elements are at the 'managed level', whereas almost all of the critical factors under the technology element are below the 'managed level'. This result concludes that the consulting organisations need significant improvements before they are ready to adopt CE. The areas which need attention and consideration are: all factors within the technology element, process focus and organisational framework within the process element and client focus within the project element.

3.4.7 Readiness of contractors

Five contracting organisations, ranging from medium-size to large, responded to the questionnaire; this represents around 40 per cent of the total number of questionnaires sent. Most of the respondents considered the people element the most important and the technology element the least important element, which is the same as for clients and consultants. The average assessment result for the contractors is plotted on the BEACON

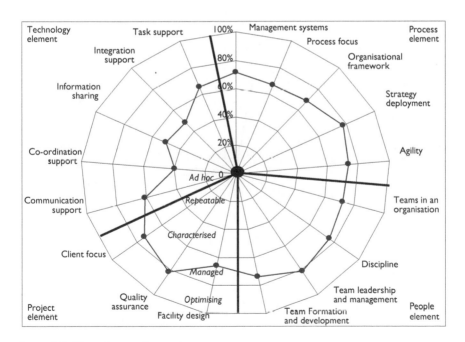

Figure 3.7 CE readiness of consultants.

model diagram shown in Figure 3.8. All the critical factors under the process and technology elements are at the 'managed level' of CE readiness whereas for the project and people elements, some of the critical factors are even at the 'optimising level'. This concludes that the contracting organisations are ready to adopt CE and have already adopted aspects of CE in some of the critical factors within the elements. The areas which need attention are communication support, and co-ordination support within the technology element, agility within the process element, teams in an organisation within the people element and client focus within the project element.

3.4.8   Readiness of sub-contractors

Twelve sub-contracting organisations, ranging from small-sized to large, were sent the BEACON questionnaire and 25 per cent of them responded. Most of the respondents commented, as did the previous groups, that the people element is the most important and the technology element the least important element from their organisational point of view. The average assessment result for sub-contractors is plotted in Figure 3.9. This shows that subcontractors are at the 'optimising level' of CE readiness except for some of the critical factors under the process, people and technology

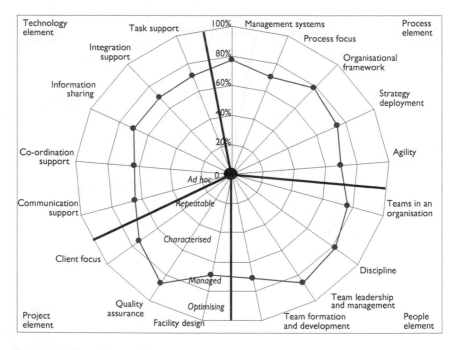

Figure 3.8 CE readiness of contractors.

elements, which are at the 'managed level'. This concludes that the sub-contracting organisations are ready to adopt CE and have already adopted aspects of CE in some areas. The areas which need to be improved are co-ordination support and information sharing within the technology element, agility within the process element and teams in an organisation within the people element.

3.4.9 Readiness of material suppliers and manufacturers

Three material suppliers and manufacturing organisations, ranging from medium to small-sized, responded to the questionnaire, which was 25 per cent of the total number of questionnaires sent. Here again, most of the organisations considered the people element as the most important and the technology element as the least important element. The readiness assessment result of the material suppliers and manufacturers is plotted on the BEACON model diagram shown in Figure 3.10. It could be seen that almost all the critical factors under the process, people and project elements are at the 'managed level' whereas all the critical factors under the technology element

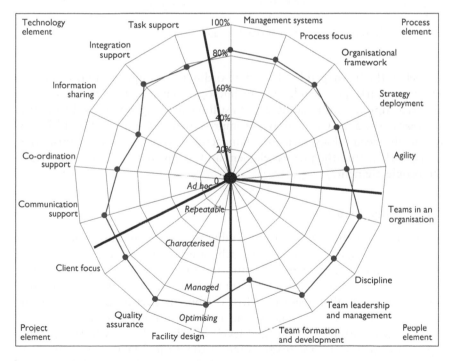

Figure 3.9 CE readiness of sub-contractors.

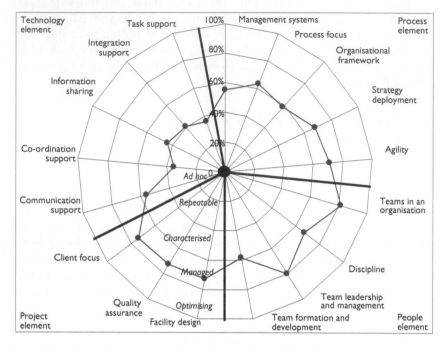

Figure 3.10 CE readiness of material suppliers and manufacturers.

are below the 'managed level'; and co-ordination support and task support are particularly poor under the technology element. This shows that the material suppliers and manufacturers still have a long way to go before they are ready to adopt CE. Significant improvements are needed in all factors within the technology element, management systems and organisational framework within the process element and teams in an organisation within the people element.

3.4.10   *Discussion*

After analysing the results of the readiness assessment case studies of the participating organisations within each category, it could be seen that the people element is considered the most important element and the technology element the least important element from most of the organisations' point of view in all categories. Most of the contracting organisations are almost ready for CE in general and most of the critical factors in each elements are within the 'managed level' of CE readiness whereas consulting organisations are not ready, some of the critical factors are within the 'managed level' while the rest are below the 'managed level' and need improvement. The same is true of client organisations, which need improvements in more than half of the critical factors within each element. The assessment results for suppliers and manufacturers portray them as the least ready for the adoption of CE. On the other hand, the results for sub-contractors show them to be the most ready for CE implementation, compared to all other sectors, with most of the critical factors at the 'optimised level'.

As far as critical factors under the process element are concerned, sub-contractors are the most advance with many factors at the 'optimising level', whereas all other sectors are at the 'managed level'. Agility is the weakest area for clients, contractors and sub-contractors, whereas process focus and management systems are the weakest areas for consultants and suppliers respectively. Material suppliers and manufacturers need the most improvements to the critical factors under the process element.

Client organisations need the most improvements within areas covered under the people element whereas sub-contracting organisations are performing well except for one factor, that is teams in an organisation, which is also the weakest critical factor in all the other sectors. Overall, for the people element, sub-contractors are at the 'optimising level' and the rest are at the 'managed level' of CE readiness.

All sectors seem to be performing well in the areas under the project element, specially sub-contractors, who are at the 'optimising level' of the CE readiness while the rest are at the 'managed level'. Client focus seems to be the weakest area for all sectors except for the client organisations, which need the most improvements within the quality assurance factor.

Critical areas covered under the technology element need the most attention and consideration by all sectors, although contractors and sub-contractors are marginally better than others, being at the 'managed level'. Clients, consultants and suppliers are all at the 'characterised level' and need considerable improvements in all areas under this element. The weakest critical factor for all sectors is co-ordination support.

The overall results show that the construction industry, as a whole still needs improvements in most of the critical areas in order to adopt CE effectively. Sectors, which seem to be ready for CE adoption are those, which are client-focused, have greater focus on monitoring and controlling of their project development process, and are continually improving their development processes and operations.

3.4.11   Benefits of the BEACON model

The development of the BEACON model is important for the implementation of CE within the construction industry as presented earlier in this chapter. The benefits of the model are outlined below:

- The BEACON model and its associated questionnaire are specifically tailored to meet the needs of the construction supply chain.

- It addresses four key elements and aspects of CE implementation which are only partly addressed by other models.
- The model will enable the development of guidelines for the effective and more appropriate implementation of CE in construction.
- The model will enable the construction industry to identify aspects of its project delivery process that require improvements to facilitate CE implementation.
- The survey and assessment could be carried out either in the form of structured interviews; or an electronic version of the questionnaire could be completed by remote respondents.
- The model is simple and easy to use. The questionnaire can be completed using tick boxes and the graphical results are automatically generated.
- Even for organisations not considering the implementation of CE, the model can act as a useful tool for self-assessment on the four key elements: technology, process, people and project.

3.5   Summary and conclusions

This chapter has discussed CE readiness assessment within the context of construction industry and presented a comparative review of the available tools and models. It has also outlined the rationale behind the development of a new CE readiness assessment model – the BEACON model – for the construction industry and presented its features. The benefits of the model and its associated questionnaire are discussed. The following conclusions can be drawn:

- Implementation of CE within the construction industry has the potential to contribute towards client satisfaction by improving quality, adding greater value, reducing cost and reducing construciton schedules.
- CE readiness assessment should be carried out before CE implementation so as to ensure that maximum benefit is achieved.
- A unique CE readiness assessment model is required for construction because existing models are not appropriate in their present form.
- The BEACON model has been developed specifically for CE readiness assessment of the construction supply chain, and will facilitate the formulation of strategies for effective CE implementation in the construction industry.

The construction industry can realise significant benefits from the adoption of CE. Readiness assessment of the industry will ensure that the right approaches are adopted for this purpose.

The assessment results show that the people element and technology element are respectively the most and least important elements for most of the organisations in all categories. Contractors and sub-contractors, in general, are ready to adopt aspects of CE within their organisations whereas clients, consultants and suppliers and manufacturers are not yet ready to adopt CE. The most important conclusion is that, overall, the construction industry is not yet ready to adopt CE and needs significant improvements in a number of critical areas before CE adoption. The industry also needs appropriate guidelines for improvements in the weaker areas as well as guidelines for the implementation of CE within the industry. Another important conclusion, which could be drawn, is that the BEACON model can be successfully used as a CE readiness assessment tool for the construction industry. It can also be used as a useful tool for self-assessment on the four key elements: technology, process, people and project for organisations CE. The work presented in this chapter is contributing in this regard and will, in future, provide detailed guidelines for the effective implementation of CE in the construction industry.

3.6 References

Ainscough, M. S. and Yazdani, B. (1999), *Concurrent Engineering within British Industry*, Proceedings of Advances in Concurrent Engineering (CE99), Bath, UK, pp. 443–448.

Al-Ashaab, A. and Molina, A. (1999), *Concurrent Engineering Framework: A Mexican Perspective*, Proceedings of Advances in Concurrent Engineering (CE99), Bath, UK, pp. 435–442.

Aouad, G., Cooper, R., Kagioglou, M., Hinks, J. and Sexton, M. (1998), *A Synchronised Process/IT Model to Support the Co-maturation of Processes in the Construction Sector*, Time Research Institute, University of Salford.

Bergman, L. and Ohlund, S. (1995), *Development of an Assessment Tool to Assist in the Implementation of Concurrent Engineering*, Proceedings of Conference on Concurrent Engineering: A Global Perspective, pp. 499–510.

Brookes, N. J., Backhouse, C. J. and Burns, N. D. (2000), *Improving Product Introduction Through Appropriate Organisation: The Development of the SIMPLOFI Positioning Tool*, Proceedings of the I MECH E Part B Journal of Engineering Manufacture, Vol. 214, No. 5, pp. 339–364.

Brooks, B. M. and Foster, S. G. (1997), *Implementing Concurrent Engineering*, Concurrent Engineering – The Agenda for Success, Medhat, S. (ed.), Research Studies Press Ltd.

Carnduff, T., Miles, J., Gray, A., Santoyridis, I. and Faulconbridge, A. (1997) *Case Adaptation and Versioning in Concurrent Engineering*, in Concurrent Engineering in Construction: Proceedings of 1st International Conference, Anumba, C. J. and Evbuomwan, N. F. O. (eds), IStructE, London, 3–4 July 1997, pp. 45–54.

CERC Report (1993), *Final Report on Readiness Assessment for Concurrent Engineering for DICE*, Submitted by: CE Research Centre, West Virginia University, June 1993.

CERC Technical Report (1992), *Process Issues in Implementing Concurrent Engineering*, Submitted by: CE Research Centre, West Virginia University, October 1992.

Chen, G. (1996), *The Organisational Management Framework for Implementation of Concurrent Engineering In the Chinese Context*, Advances in Concurrent Engineering, Proceedings of 3rd ISPE International Conference on Concurrent Engineering: Research and Applications, University of Toronto, Ontario, Canada, 26–28 August 1996, pp. 165–171.

Componation P. J. and Byrd Jr, J. (1996), *A Readiness Assessment Methodology for Implementing Concurrent Engineering*, Advances in Concurrent Engineering, Proceedings of 3rd ISPE International Conference on Concurrent Engineering: Research and Applications, University of Toronto, Ontario, Canada, 26–28 August 1996, pp. 150–156.

CONCUR (1999), Web Page of the CONCUR Project: http://ivope.ivo.fi/concur

Crow, K. A. (1994), *Building Effective Product Development Teams*, DRM Associates, 1994.

de Graaf, R. and Sol, E. J. (1994), *Assessing Europe's Readiness for Concurrent Engineering*, Proceedings of Conf. on Concurrent Engineering: Research and Application, 1994, pp. 77–82.

Duke, A. K. and Anumba, C. J. (1997), *Enabling Collaboration for Virtual Construction Project Teams*, in Concurrent Engineering in Construction: Proceedings of 1st International Conference, Anumba, C. J. and Evbuomwan, N. F. O. (eds), IStructE, London, 3–4 July 1997, pp. 163–172.

Egan, J. (1998), *Rethinking Construction*, Report of the Construction Task Force on the Scope for Improving the Quality and Efficiency of UK Construction Industry, Department of the Environment, Transport and the Regions, London.

Evbuomwan, N. F. O. and Anumba, C. J. (1998), *An Integrated Framework for Concurrent Life-cycle Design and Construction*, Advances in Engineering Software, 1998, Vol. 5, Nos. 7–9, pp. 587–597.

Finnemore, M. and Sarshar, M. (2000), *Linking Construction Process Improvement to Business Benefit*, Bizarre Fruit 2000 Conference, University of Salford, 9–10 March 2000, pp. 94–106.

Kamara, J. M., Anumba, C. J. and Evbuomwan, N. F. O. (2000), Developments in the Implementation of Concurrent Engineering in Construction, *International Journal of Computer Integrated Design and Construction*, Vol. 2, No. 1, pp. 68–78.

Khalfan, M. M. A. (2000), *Improving Business Performance in Construction through Integration and Concurrent Engineering*, Business Information Systems 2000 Conference, Poznan, Poland, 12–13 April 2000, see web site: http://www.bis.pozn.com

Khalfan, M. M. A. and Anumba, C. J. (2000a), *Readiness Assessment for Concurrent Engineering in Construction*, Bizarre Fruit 2000 Conference, University of Salford, UK, 9–10 March 2000, pp. 42–54.

Khalfan, M. M. A. and Anumba, C. J. (2000b), *Implementation of Concurrent Engineering in Construction – Readiness Assessment*, Construction Information Technology 2000 Conference, Reykjavik, Iceland, 28–30 June 2000, Vol. 1, pp. 544–555.

Khalfan, M. M. A. and Anumba, C. J. (2000c), *A Comparative Review of Concurrent Engineering Readiness Assessment Tools and Models*, Concurrent Engineering 2000 Conference, Lyon, France, 17–20 July 2000, pp. 578–585.

Khalfan, M. M. A., Anumba, C. J., Siemieniuch, C. E. and Sinclair, M. A. (2001), Readiness Assessment of the Construction Supply Chain for Concurrent Engineering, *European Journal of Purchasing and Supply Management*, Vol. 7, Issue 2, pp. 141–153.

Kwak, Y. H. and Ibbs, C. W. (1997), *Project Management Process Maturity (PM)^2 Model*, see web site: http://www.ce.berkeley.edu/~yhkwak/pmmaturity.html

Latham, M. (1994), *Constructing the Team*, Final Report on Joint Review of Procurement and Contractual Agreements in the UK Construction Industry, HMSO, London.

Love, Peter E. D. and Gunasekaran, A. (1997), *Concurrent Engineering in the Construction Industry*, Concurrent Engineering: Research & Applications, Vol. 5, No. 2, pp. 155–162.

Martin, A. and Evans S. (1992), *Project Planning in a Concurrent Engineering Environment*, Proceedings of 3rd International Conference on Factory 2000, New York, UK, July 1992, pp. 298–303.

Muya, M., Price, A. D. F. and Thrope, A. (1999), *Contractors' Supplier Management*, Proceedings of a Joint Triennial Symposium, Cape Town, 5–10 September 1999, Vol. 2, pp. 632–640.

Paul, J. and Burns, C. (1997), *Implementing a Business Process Re-engineering Programme*, Concurrent Engineering – The Agenda for Success, Medhat, S. (ed.), Research Studies Press Ltd.

Prasad, B., 1997, *Seven Enabling Principles of Concurrency and Simultaneity in Concurrent Engineering*, Concurrent Engineering in Construction, Anumba, C. J. and Evbuomwan, N. F. O. (eds), Proceedings of 1st International Conference organised by The Institution of Structural Engineers Informal Study Group on Computing in Structural Engineering, London, 3–4 July 1997, pp. 1–12.

Rezgui, Y., Cooper, G., Yip, J. and Brandon, P. (1997), *Construction Collaborative Engineering in a Concurrent Multi-actor Environment*, in Concurrent Engineering in Construction: Proceedings of 1st International Conference, Anumba, C. J. and Evbuomwan, N. F. O. (eds), IStructE, London, 3–4 July 1997, pp. 183–194.

SPICE Questionnaire (1998), *Key Construction Process Questionnaire*, Ver. 1.0, Salford University, July 1998.

ToCEE (1997), *ToCEE European Project for Concurrent Engineering*, ToCEE Information Document, 1997.

Wailes, M. K., Tizani, W., Nethercot, D. A. and Smith, N. J. (1997), *The Integration of Tools for Concurrent Design and Fabrication of Steel Structures*, in Concurrent Engineering in Construction: Proceedings of 1st International Conference, Anumba, C. J. and Evbuomwan, N. F. O. (eds), IStructE, London, 3–4 July 1997, pp. 100–108.

Winner, R. I., Pennell, J. P., Bertrend, H. F. and Slusarczuk, M. M. G. (1988), *The Role of Concurrent Engineering in Weapons System Acquisition*, IDA Report R-338, Institute for Defence Analyses, Alexandria, VA.

Wognum, P. M., Stoeten, B. J. B., Kerkhof, M. and de Graaf, R. (1996), *PMO-RACE: A Combined Method for Assessing Organisations for CE*, Advances in Concurrent Engineering, Proceedings of 3rd ISPE International Conference on Concurrent Engineering: Research and Applications, University of Toronto, Ontario, Canada, 26–28 August 1996, pp. 113–120.

Yin, R. K. (1989), *Case Study Research: Design and Methods*, Applied Social Research Methods Series, Vol. 5, Sage Publications, 1989.

Young, R. I. M. (1999), *Organisational Issues in Concurrent Engineering*, Department of Manufacturing Engineering, Loughborough University, Lecture Notes, Section 2, pp. 8–15.

Chapter 4

The 'voice of the client' within a Concurrent Engineering design context

John M. Kamara and Chimay J. Anumba

4.1 Introduction

Clients are crucial to the construction process, since they initiate and pay for projects. Although they were traditionally not considered to be part of the construction industry (Fellows *et al.*, 1983; Marks *et al.*, 1985), their active involvement is crucial for project success, and is also considered to be a key strategy for accelerating change in the industry (Kometa *et al.*, 1995; SFC, 2002; CCC, 2004). The Clients' Charter of the Confederation of Construction Clients (CCC) in the UK makes it clear that clients should provide effective leadership of the construction process 'through making their main project requirements fully transparent and creating the right environment for the supply-side to meet those requirements in the most effective way' (SFC, 2002).

The active involvement of clients in the procurement of their facilities suggests that clients are now part of the project process, or that design and construction professionals should now be involved in 'pre-project' activities such as the establishment of the need to build. This idea is reinforced by earlier research into the development of Generic Design and Construction Processes (otherwise known as the Process Protocol) (see Chapter 3 in this book), and more recent calls for the adoption of long-term framework agreements by clients (CCC, 2004).

Figure 4.1 shows a conceptual model of a Concurrent Engineering (CE)-based design context, which shows the schematic integration of the functional disciplines involved in a project, the design process and design tools, and the integration of textual and geometric project data (Evbuomwan and Anumba, 1995). It consists of three levels: design stages, design tools and techniques, and knowledge bases and databases. Level 1 consists of the following stages: preliminary or concept design, scheme design, detailed design, design documentation and construction planning. Information on client requirements feeds into the design process, but can also be part of the Level 1 activities. Level 2 represents computer-aided design tools, as well as other design methods and techniques

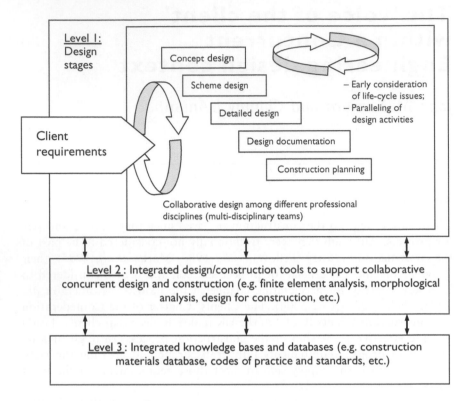

Figure 4.1 A CE project model.

Source: Kamara *et al.* (2000).

including: codes of practice and industry standards that can be used in performing design activities at any of the design stages shown in Level 1. Level 3 consists of the necessary knowledge and databases that support the design tools represented in Level 2.

The idea of a joined-up project process that includes the client provides a much more stable environment for the implementation of CE. But it is also necessary to specify how the wishes of clients can be elicited and presented in a format that facilitates concurrent working and the implementation of client requirements. This chapter describes an innovative approach (the Client Requirements Processing Model – CRPM) for encapsulating the 'voice of the client' within a CE-based design context. The context and rationale for the development of the CRPM, and its key features, are described. The chapter concludes with an illustrative example of how the CRPM can be used in practice, and the ways in which it can contribute to the implementation of CE in construction projects.

4.2   Client requirements processing (CRP)

A client can be defined as the person or organisation responsible for commissioning and paying for the design and construction of a facility (e.g. building, road, bridge), and is usually (but not always) the owner of the facility being commissioned. The client can also be the user of a proposed facility, or they (i.e. client and user) may be separate entities. However, as the purchaser of services for the design and construction of a facility, the client represents (and should consider) other interests. These include the owner, if different, users and other identified persons, groups or organisations who influence, and are affected by the acquisition, use, operation and demolition of the proposed facility (e.g. financial institutions, environmental pressure groups and neighbourhood associations). Thus the 'client' (i.e. buyer of construction services) is a 'body' or 'entity' that incorporates other interests groups. The extent to which these are involved depends on the kind, and scale of the project. A road project and/or a nuclear power station, for example, will attract the attention of environmental groups; the citing of an entertainment facility such as a nightclub, in a residential area will have to consider the views of the residents of that neighbourhood (Kamara *et al.*, 2002).

Client requirements refer to the objectives, needs, wishes and expectations of the client. These requirements should also be a description, with respect to functions, attributes or other special features, of the facility that satisfies the client's objectives (or business need). Client requirements constitute the primary source of information for a construction project, and are therefore of vital importance to the implementation of CE in construction.

The expression of the needs of a client in a form that describes the facility that he/she desires involves some form of 'processing'. Where the client is likely to express his/her needs in non-design terms it then becomes necessary to 'translate' them into design terms. CRP therefore, involves the presentation of information in a format that enhances the understanding of precisely what the client desires. It can be defined as the identification (or definition), analysis, and translation of explicit and implicit client requirements into solution-neutral design specifications (Kamara *et al.*, 2002).

CRP can be seen both as an input to design and construction, and as part of an integrated design and construction framework (Sanvido and Norton, 1994; Evbuomwan and Anumba, 1995). As an input to design, it provides an interface between a client's demands and the measures (design and construction) used by the industry to meet those demands (Worthington, 1994; Gibson *et al.*, 1995). As part of the design and construction of a facility, CRP is incorporated in the integrated CE project model shown in Figure 4.1. Therefore, CRP should both facilitate 'concurrent' working (as an input to the design and construction process), and reflect the philosophy of CE (as part of an integrated CE process). These two aspects of the requirements for CRP in CE are described later.

4.2.1   Facilitating 'concurrent' working

For the client requirements processing activity to facilitate concurrent working, the outputs of the process should be in a format that will enable:

- different disciplines to work concurrently (or in parallel) as much as possible;
- the early (up-front) consideration of all life-cycle issues affecting the facility;
- the integration of all the professional disciplines involved in the process;
- the traceability of design decisions to original requirements throughout the life-cycle of the facility.

This, in effect, deals with the nature and content of the information generated from the requirements processing activity. Thus, the focus is on how the client requirements are 'expressed' or 'stated'. This suggests that, for different disciplines to work concurrently, they should be able to have the same set of requirements, and fully understand their meaning from the perspective and priorities of the client. To facilitate concurrent working therefore, the requirements of the client should be:

- precisely defined to remove any ambiguities;
- stated in a solution-neutral format that can be understood by the different disciplines working on a project;
- stated in a format which makes it easy to trace and correlate design decisions to the original intentions of the client (Perkinson *et al.*, 1994);
- reflective of all the perspectives and priorities of the client (owner, user and other 'interests').

4.2.2   Compatibility with the CE philosophy

This deals with the framework for requirements processing (i.e. the way in which it is carried out) which should ensure that the outputs of the process are in line with those stated earlier. However, because CRP is part of an integrated CE framework, it should reflect the principles of CE. Thus, it should be characterised by the CE principles discussed earlier: the participation of multi-disciplinary teams, early or up-front consideration of life-cycle issues, integrated teamwork where activities are carried out in parallel, continuous improvement by incorporation of lessons learned where possible, and continued focus on the requirements of the client. It is also essential that a framework for CRP in a CE context is computer-based, in order to realise the full benefits of CE (Kuan, 1995; Prasad, 1996). A computer-based CRP framework is vital for integration with IT-based downstream activities in construction. The use and implementation of structured methodologies are also better managed using IT tools. Furthermore, conformance checking and traceability of requirements throughout the project life-cycle can be automatically done if the processing of requirements in CE is computer-based.

4.2.3   Client and project requirements

Another reason why it is necessary to 'process' client requirements is that, within the context of the project in which they are implemented, there are other requirements (Table 4.1).

Client requirements combine with site, environmental and regulatory requirements to produce design requirements, which in turn generate construction requirements. Other project requirements are generated

Table 4.1 The different requirements represented in a project

Type of requirements	Meaning
Client requirements	Requirements of the client that describe the facility that satisfies his or her business need. Incorporates user requirements, and those of other interest groups
Site requirements	These describe the characteristics of the site on which the facility is to be built (e.g. ground conditions, existing services, history, etc.)
Environmental requirements	These describe the immediate environment (climatic factors, neighbourhood, etc.) surrounding the proposed site for the facility
Regulatory requirements	Building, planning, health and safety regulations, and other legal requirements that influence the acquisition, existence, operation and demolition of the facility
Design requirements	These are the requirements for design which are a translation of the client needs, site and environmental requirements. They are expressed in a format that designers can understand and act upon
Construction requirements	These are the requirements for actual construction which follow from the design activity
Life-cycle requirements (LCR)	These go beyond project completion and include the requirements for operating and maintaining the facility, its disposal or recycling. LCRs are strictly not project requirements, but as a construction project is not an end in itself, it is necessary that they are considered during the Project, preferably within the client requirements

(or derive) from the business need of the client that is to be satisfied by the proposed facility. For example, a client's desire to have an office block in a strategic location (because of the nature of his or her business activities) will have an effect on the site, environmental and regulatory (relevant planning regulations) requirements. This suggests that, other project requirements can either pose constraints to client requirements, or they can enhance their satisfaction. An adequate understanding of client requirements (through effective processing) can therefore facilitate the level of trade-offs required with other project requirements, which are more difficult to alter than client requirements.

4.3   CRP and project briefing

The process for eliciting and defining client requirements in construction is referred to as briefing (or facility programming – Perkinson *et al.*, 1994). The document, which contains these requirements, is referred to as the brief (or 'programme'). A review of existing literature on briefing (e.g. Parsloe, 1990; Salisbury, 1990; CIB, 1997), case studies, discussions with construction professionals and clients, and a structured postal questionnaire survey were used to assess how briefing is carried out in the UK construction industry (Kamara and Anumba, 2001). The main features of the briefing process (summarised in Table 4.2) include: the dominance of design professionals in briefing, the consideration of briefing and design as one activity, and the use of sketches and drawings (design) to clarify the client's objectives.

4.3.1   Limitations

The findings from case studies and the survey provided an insight into the problems in current briefing practice (Kamara and Anumba, 2001). They include the following:

- inadequate involvement of all the relevant parties to a project;
- insufficient time allocated for briefing;
- inadequate considerations of the perspectives of the client;
- inadequate communication between those involved in briefing;
- inadequate management of changes to requirements.

These problems, which are supported by other studies on briefing (e.g. Newman *et al.*, 1981; Goodacre *et al.*, 1982; CIT, 1996), may be due to the attitude or inefficiencies of those involved, but they also suggest that the general framework for briefing is inadequate.

Table 4.2 Findings about the briefing process

Briefing process	Findings
Those involved in briefing	A broad mix of professionals (both within and outside the client organisation) are involved in briefing; they include: administrators (managers), architects, development managers, engineers (building services, civil, structural), planning supervisors, portfolio managers, project managers, quantity surveyors (QS), etc.; design professionals (e.g. architects however, tend to dominate the briefing process
Stages in briefing	Briefing is combined with design (i.e. conceptual and scheme design), and usually, there are no distinct stages in the process; briefing information becomes more detailed as design progresses
Collection and documentation of information	A variety of methods are used to collect information: for example, interviews, workshops, evaluation of existing facilities, visits to similar facilities, etc.; information collected is sometimes documented in formal documents (e.g. letters, faxes, e-mail, minutes of meetings, sketches and drawings, etc.); these documents are not normally stored as part of 'the brief', and usually, design team relies on recollections of verbal communications with the client
Processing of information	A process of 'trial and error', through the use of sketches and drawings, is mostly used to clarify the client's problem, or process briefing information; there are situations, however, where clients who commission many projects, define their requirements before design
Decision-making in briefing	Decision-making involves the resolution of competing interests between different groups within the client body, and between professionals with diverse perspectives; decisions are usually the result of discussions and negotiations between those involved; techniques such as value management are used to assist in decision-making
Management of the briefing process	Management of changes to requirements is influenced by the way requirements are represented in subsequent stages of the briefing and design process; changes to requirements are managed by recording them as corrections to sketches and drawings, the main medium for representing the brief; changes may also be discussed in meetings and decisions recorded in the reports (minutes) of those meetings

Source: Kamara and Anumba (2001).

4.3.2 Limitations in the framework for briefing

Current briefing practice deals with the collection of information for project implementation, and often, project requirements are taken to be the same as client requirements. However, as discussed previously, an adequate understanding of client requirements can only be achieved if they are considered distinctly from other project requirements, so that the problem that design and construction are to solve, within the context of the site and immediate environment, can be clearly defined (ensuring that 'the tail doesn't wag the dog').

Another limitation is that, use of the solution (i.e. design) to clarify the problem, can also shift focus from client requirements to the preferences of designers. This is because, proposed solutions are usually made before a thorough understanding of the client's requirements. There is therefore an inherent tendency for the client to be influenced by the preferences of the designer(s). This in itself may not be disadvantageous to the client, who relies on the expertise of the designer to provide a design solution to his or her problem. However, as MacLeod *et al.* (1998) put it, 'if one does not know clearly what one is trying to achieve...then the chances of achieving good outcomes must be diminished'. Furthermore, this practice assumes that a design professional has to lead the briefing process. But designers are not necessarily good brief writers since briefing is mainly concerned with the processing of information (Palmer, 1981). It is therefore not surprising that many briefs are generated out of design rather than a clear understanding of the client's actual objectives (Howie, 1996).

4.3.3   Need for an effective framework for CRP

The limitations in the process and framework for briefing have led to the realisation that, focus on the client's business need, and the use of structured methods, to facilitate the definition, analysis, documentation, traceability and correlation of all relevant information, can enhance the briefing process (Newman *et al.*, 1981; Farbstein, 1993; Worthington, 1994). To this end, various research initiatives have been undertaken to devise ways to improve the briefing process. These include the development of computer and information tools to assist in the creation and management of briefing information, and the use of techniques from manufacturing to analyse client client requirements (Mallon and Mulligan, 1993; Perkinson *et al.*, 1994; Kumar, 1996; Serpell and Wagner, 1997; Yusuf, 1997, Rezgui *et al.*, 2003; Othman *et al.*, 2004). However, these efforts do not adequately provide for the effective processing of client requirements within a CE-based project environment. For example, those that are based on the development of software to support briefing are basically computerised systems of existing practices without any re-engineering of the process. There is also no comprehensive framework to incorporate and prioritise the different perspectives represented by the client. It is therefore evident that an effective framework for the processing of client requirements for CE is needed. This is provided for by the CRPM.

4.4   The CRPM

The CRPM describes the methodology for processing client requirements within a CE environment. It is represented using the Integration Definition method for functional modeling (IDEF-0) and the EXPRESS-G graphical notation for information modeling (IDEF, 1993; Schenck and Wilson, 1994). The rationale for using these modeling methods is based on their relatively ease of use and understanding, and the fact that they have been proven to be appropriate in construction (Sanvido *et al.*, 1990; Hannus, 1992, Vanier *et al.*, 1996; Karhu *et al.*, 1997, Yusuf, 1997). Furthermore information models described using EXPRESS-G are independent of any implementation context, allowing flexibility in the computer implementation of the model.

4.4.1   The main stages of the CRPM

Figure 4.2 shows the context diagram for the CRPM showing the three stages of the model: 'define client requirements', 'analyse client requirements' and 'translate client requirements'. The details of the activities in each stage are summarised in Table 4.3. 'Define requirements' deal with the identification of interest groups represented by the client, and the elicitation of requirements. The 'analyse client requirements' stage deals with the structuring and prioritisation of client requirements based on the relative importance interest groups place on those requirements. The 'translate client requirements' stage deals with the generation and prioritisation of design attributes, calculation of target values, translation of client requirements into design attributes.

Design attributes are 'metrics' which can be used to 'measure' (or assess) something. Within the context of building/construction design, a design metric is made up of three parts: the metric, a 'measurement' scale and a 'target value'. The 'metric' is a statement of what is to be measured (e.g. gross floor area). The 'measurement' scale refers to a clearly defined means of assessing the metric. This can include a unit (e.g. m^2) or statement describing how the metric will be assessed (e.g. 'number of unplanned repairs per year'). The 'target' value refers to the range (or design solution space) within which a designer should operate to achieve the client's objectives (e.g. 'between 50 and 100 m^2'). An example of a design metric, and how it relates both to client requirements and design solutions, is illustrated in Figure 4.3.

When client requirements and design metrics are considered, a client requirement is an independent variable and a design metric is a dependent variable (i.e. a client requirement produces a change in how much space is to be designed for). On the other hand, when design metrics and design solutions are considered, the design metric becomes the independent

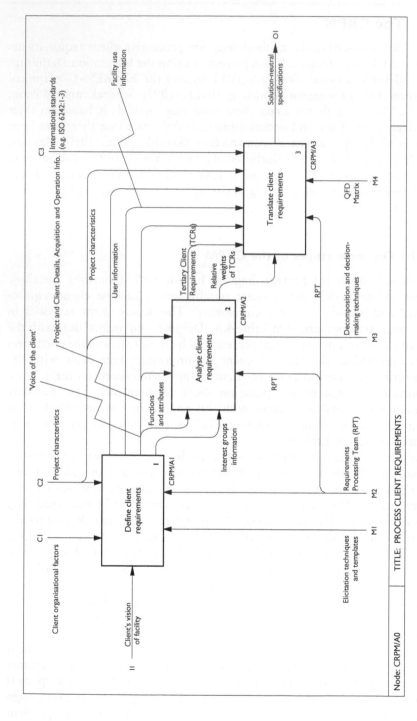

Node: CRPM/A0   TITLE: PROCESS CLIENT REQUIREMENTS

Figure 4.2 Context diagram showing the three main stages of the CRPM.

Table 4.3 The main stages and activities of the CRPM

Main stage	Activities	Required resources (tools)
Define client requirements	• Establish and document basic facts about the project and the client; • Identify and describe the people or groups ('interest groups') which influence, and/or are affected by the acquisition, operation/use and existence of the proposed facility; • Elicit from client, the functions and attributes of the proposed facility, information on its acquisition, operation, future demolition, activities to be performed in the facility, and the characteristics of proposed users ('voice of the client')	• A multi-disciplinary requirements processing team; • Elicitation techniques (e.g. questionnaires, interview techniques)
Analyse client requirements	• Structure and prioritise client requirements; • Restate (or decompose) client requirements into primary, secondary and tertiary requirements to facilitate a clearer understanding of those requirements; • Determine the relative importance of 'interest groups'; • Prioritise tertiary requirements with respect to the relative importance of each interest group and their weighting of each tertiary requirement	• Requirements processing team; • Value tree analysis to decompose requirements; • Decision-making techniques (e.g. criteria weighting)
Translate client requirements	• Generate design attributes; • Determine target values for these design attributes using information on the characteristics of the project, proposed use and users of the facility, acquisition and operation of the facility, international standards (including codes of practice), and target values for similar facilities); • Translate tertiary client requirements by matching them with identified design attributes to determine which design attributes best satisfy a particular requirement; • Prioritise design attributes which have been matched with client requirements • Prioritised design attributes and their target values constitute the solution-neutral specifications	• Requirements processing team; • The QFD 'house of quality' matrix

Source: Kamara and Anumba (2000).

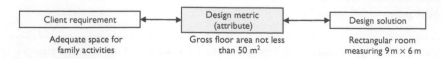

Figure 4.3 Client requirement, design metric and design solution.

variable, and the design solution the dependent variable (i.e. a design metric influences the kind of solution produced, but is itself not part of the solution). The concept of design metrics is widely used in product design, where the focus tends to be quantitative. Even in building design, quantitative factors (e.g. heating and lighting levels, etc.) can be easily represented in the form of design metrics. The challenge is however the more qualitative (or architectural) aspects of design. For example, if a client requirement is for 'adequate security' how does one objectively assess whether this feature has been achieved in a design?

4.4.2   Informational representation of the CRPM

Figure 4.4 shows the EXPRESS-G representation of the CRPM. Client requirements, expressed as primary, secondary and tertiary requirements (with absolute and relative weights), describe the facility that satisfies the business need of the client.

The requirements of the client consists of information relating to: the characteristics or nature of the client and the project ('client/project characteristics'), his or her business need ('client business need'), and the acquisition, operation and disposal of the facility ('facility "process"'). Client requirements are influenced by 'other sources of information' in the sense that a change in some standards (e.g. space standards or energy emission targets) might influence the decision by a client to commission the refurbishment of an existing building. The client organisation determines the business need for a project. On the other hand, the business need (e.g. improved communication between two locations), influences the type of project (e.g. refurbishment) as well as the interest groups associated with the process and outcome of that project – the facility. The nature of the client organisation, and the kind of project, will also determine how the facility is procured, operated and disposed of. For example, a client organisation with a substantial property portfolio can have personnel who are responsible for the acquisition of new property, unlike a one-off client who might require considerable assistance from outside consultants. 'Facility "process"' on the other hand, will have an influence on the organisation of the project. The entity, 'solution-neutral specifications', shown in Figure 4.4,

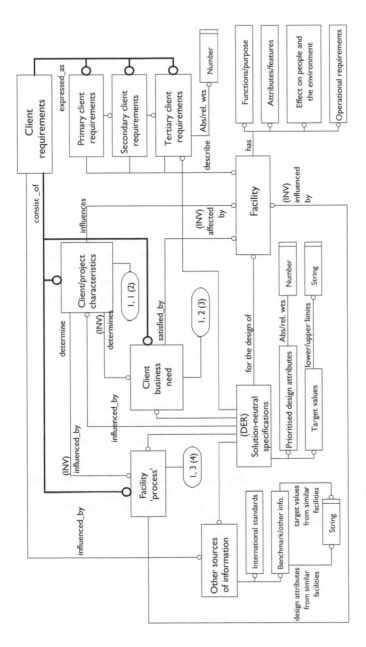

Figure 4.4 Complete entity diagram of the CRPM information model.

is derived from 'other sources of information', 'facility "process"', 'client/ project characteristics', 'client business need' and 'tertiary client requirements'. Solution-neutral specifications, required for the design of the facility that satisfies the business need of the client, consist of prioritised design attributes (absolute and relative weights) and target values. The entities and attributes for the CRPM are listed in Table 4.4.

Table 4.4 Attributes for defined entities in the CRPM

Entity group	Entity	Attributes
Client/project characteristics	Client details	Name, address, business type, contact person name, contact person address, number of employees, average annual turnover, client policy on occupancy and space utilisation
	Interest groups	Group name, type of group, relationship with client, group's influence in acquisition, operation and use of facility, effect of facility acquisition, operation and use, on group
	Project details	Project name, project location, project type, facility type, facility objectives
Client business need	Facility use information	Activity type, time of day performed, time of year performed, peak use times, required equipment and furniture
	User information	User name, user type, user size, relationship with client, activity user performs
	Facility functions	Function verb, function noun, function qualifier, functions rationale (i.e. why a specific function is required)
	Facility attributes	Attribute name, attribute meaning, attribute rationale, function associated with attribute
Facility 'process'	Acquisition information	Available budget, rationale for budget allocation, level of client involvement (Risk), rationale for level of client involvement, approved client representatives, expected date of completion, Rationale for completion date
	Operation information	Costs in use, meaning and rationale for costs in use, operation/management strategy, rationale for operation/management strategy, level of operation/management technology, rationale for operation/management technology
	Disposal information	Expected life span, rationale for expected life span, etc.
Other sources of information	International standards	Standards for the expression of user requirements, standards for air capacity for occupants in specified building types, etc.
	Benchmark/ other information	Operation/maintenance information for existing or similar facilities, etc.

4.5   Practical application of the CRPM

An illustrative example on the use of the CRPM is now described. This provides the basis for discussing how the CRPM facilitates the incorporation of the 'voice of the client' within a CE-based project process. The example used here is that for a hypothetical road project, which is based on the retrospective use of the CRPM on the Newbury Bypass Project (Kamara *et al.*, 2002). It is assumed here that a section of the A1 in Northumberland (UK) is being diverted to bypass a small market town (referred to here as 'Bobton') and thereby alleviate traffic congestion through the town.

4.5.1   Define client requirements

A key to the 'define client requirements' activity for a project like this is the identification of the components of the client and those who will influence, and be affected by the acquisition, operation and disposal of the facility. These interest groups include: the highways agency, other government departments, local government authority, local residents, politicians (both local and national), motorists, environmental and other pressure groups (e.g. the roads lobby, Friends of the Earth, etc.), and relevant statutory bodies. Drawing from the Newbury Bypass example (Kamara *et al.*, 2002), the functions and attributed of the 'facility' can be defined as in Table 4.5.

Other information that should be captured at this stage includes acquisition information (allowable budget, expected duration, appointed representatives of the client, etc.), operation information (whether a toll system is preferred, effective speed management, etc.), and disposal information (e.g. expected life-span, future plans, etc.).

4.5.2   Requirements analysis

This involves the structuring of requirements into primary, secondary and tertiary requirements (Table 4.3), the prioritisation of interest groups and the prioritisation of tertiary requirements, using the techniques such as Value Tree Analysis and Criteria Weighting (Kamara *et al.*, 2002).

An examination of the list of functions and attributes for the road in Table 4.5 reveals that there is repeated reference to the safety and quality of life for local residents. This suggests that a strategic need (primary requirement) for the project is the improvement of the quality of life for local residents. Another strategic need could be the need to implement government plans for road building (Figure 4.5).

With regard to the prioritisation of interest groups and tertiary requirements, the relative importance of each interest group should be assessed

Table 4.5 Functions and attributes of 'Bobton' bypass road project

Functions/attributes	Rationale
Functions: The road should	
• Provide a single direct route that stretches the length of the country	There is a need for through traffic on the A1 up to Edinburgh in the UK
• Reduce the amount of traffic passing through the centre of 'Bobton'	Traffic through the town centre causes congestion and traffic jams, and poses a risk to local residents
	The bypass will divert non-local traffic away from the town
• Reduce journey times	The reduction of journey times through less congestion will facilitate the smooth operation of business activities
• Reduce pollution	This is necessary to increase the quality of life of local residents
• Minimise the disruption to local traffic and pedestrian movement	Since the A1 goes through the town, out-of-town traffic causes congestion and disruption to local traffic
• Improve the quality of life of citizens in the town	Increased traffic causes pollution through noise and the discharge of exhaust fumes. The vibration from heavy traffic also causes damage to structures
Attributes: The road should be	
• Environmentally sustainable (i.e. environmental issues and concerns duly taken into consideration)	Protection of the natural and built environmental is at and concerns are the heart of Government policy
• Cost-effective to construct (i.e. the most economical solution should be pursued)	The government seeks to be prudent in its use of public funds

Source: Kamara *et al.* (2002).

with respect to their influence on the acquisition, use, and operation of the facility, and on the effect of the road on them. The extensive consultation process and public enquiries for road projects provides a good opportunity to assess the relative weightings different interest groups place on various requirements.

4.5.3   Requirements translation

Table 4.6 provides a list of design attributes and suggested target values, which were compiled from a Highways Agency document on route management strategies (Highways Agency, 2000). In generating this list, attention was paid to the requirements for the project, to ensure that the technical specifications (design attributes) selected will satisfy the tertiary

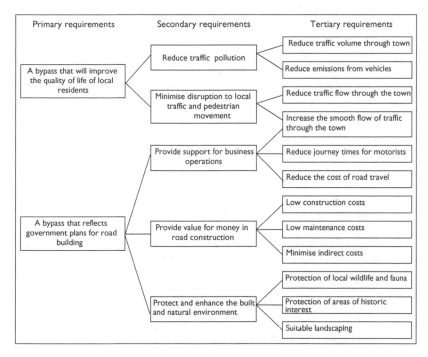

Figure 4.5 Primary, secondary and tertiary requirements for the 'Bobton' bypass project.
Source: Kamara *et al.* (2002).

requirements. For example, the following tertiary requirements (Figure 4.5): 'reduce traffic volume through town', 'reduce journey times for motorists', 'reduce the cost of travel', etc. could be satisfied, to varying degrees, by the design attribute 'minimum traffic congestion', which deals with the congestion delays. Similarly, 'protection of local wildlife and fauna' and 'suitable landscaping', could be satisfied by the design attributes, 'adequate biodiversity action plan' and 'appropriate land management plans'.

The design attributes, are technical performance (solution-neutral) specifications that have implications for design. For example, to satisfy the 'minimum traffic congestion' target, consideration will have to be given to project traffic volumes, physical properties of the road (e.g. number of carriageways), and the traffic management scheme for the road (e.g. speed limits). Similarly, the 'minimum noise exposure' design attribute will affect the design of road surfacing in the vicinity of properties, location of embankments and cuttings, etc. Thus, by systematically mapping tertiary requirements (which relate to strategic business needs) to design attributes,

Table 4.6 Sample design attributes for the 'Bobton' bypass road project

Notation	Design attribute	Unit of measurement	Target value
D1	Minimum traffic congestion	Peak average time delay per vehicle km	Less than 10 secs
D2	Low accident rates	Personal Injury Accidents (PIA) per 100 veh. kms	Not more than 15
D3	Minimum noise exposure	Noise severity index expressed number of properties per km where people are expected level above 68 dBA	Less than 10
D4	Adequate biodiversity action plan	% of road section for which a plan exists	Greater than 75%
D5	Appropriate land management plans	% of road section for which a plan exists	Greater than 75%
D6	Minimum hindrance to non-motorised road users (e.g. number of at grade pedestrian crossings)	A score reflecting the need of users and quality of facilities on a scale of 5 (highest) to 1 (lowest hindrance to non-motorised road users)	1
D7	Adequate road facilities (e.g. emergency telephones, lay-bys)	Availability of facilities score on a scale of 5 (least facilities) to 1 (most facilities)	1
D8	Road capacity (i.e. traffic volume on new bypass)	Proportion of traffic volumes passing through town	At least 50%
D9	Flexibility to handle future growth	Projected traffic (PT) volumes over the design life	Road should handle at least 90% of PT

Source: Kamara *et al.* (2002).

proposed designs can easily be checked to see if they incorporate the 'voice of the client'.

4.6   The CRPM and a CE design context

The CRPM was developed to facilitate the incorporation of the 'voice of the client' in a CE-based project context in construction. The methodology it represents is structured and focuses on the description of the proposed facility that satisfies the business need of the client. The description is not based on the physical components of the facility (e.g. shape, materials, etc.), but on its functions, attributes, acquisition, operation and effect on people and the environment (Table 4.5). The manner in which the requirements for CRP in CE are satisfied by the model is now presented.

4.6.1 Precise definition of requirements

The define requirements function provides for the precise establishment of the wishes and expectations of the client (and the different interests it represents). From the informational perspective of the model (Table 5.4), it should be noted that information is solicited on the rationale for certain statements and desires of the client. This assists the requirements processing team (RPT) to further clarify the real intentions of the client. It also ensures that stated functions and attributes are not just 'wish lists', but are based on the descriptions that reflect the real (business) needs of the client. The structuring of requirements into primary, secondary and tertiary requirements further helps in clarifying and stating requirements in a concise and unambiguous manner. It also facilitates the tracing of requirements to the original intentions of the client. It must be emphasised that the kind of information elicited should generally focus on articulated needs. There are three categories of client expectations: basic, articulated and exciting needs (Kamara et al., 2002). Basic needs are those which are not voiced but are assumed to be present in a facility (e.g. the expectation the road, used in the example earlier, is structurally sound). The fulfilment of basic needs will not excite a client, but their omission will reduce his/her satisfaction. Articulated needs are those which are voiced of demanded (e.g. reduce journey times in Table 4.5). Exciting needs are those which, although not voiced, will pleasantly surprise the client if fulfilled. Although all there categories of needs have to be fulfilled to satisfy the client, the focus should be on articulated needs because a thorough 'processing' of these needs will lead to the discovery of 'cxciting' needs.

4.6.2 Reflective of the perspectives and prioritise of the client

The need to reflect the perspectives and priorities represented by the client is addressed by the information on interest groups, the prioritisation of these groups, and the prioritisation of tertiary requirements. The information on interest groups not only identifies these groups, but also specifies how they influence, or are affected by the acquisition, operation and existence of the facility. A systematic process which incorporates the preferences of interest groups, is adopted for the prioritisation of client requirements. Furthermore, the use of formal decision-making techniques minimises bias in decision-making, but does not altogether remove the skills and experience of the RPT involved in the process. The RPT is key to the utilisation of the CRPM. It is therefore necessary that it is a multi-disciplinary team of the major disciplines involved in the lifecycle of a facility (e.g. architects, contractors, development managers, engineers, facilities managers, QC, etc.) depending on the size and nature of the project. Members of the RPT should also have sufficient knowledge of the construction process and the client organisation to make the required value judgements involved in implementing the CRPM.

4.6.3   Translation and presentation of requirements in a solution-neutral format

The requirements definition process focuses on the description of the proposed facility using terminology that is familiar to the client. The translation of requirements into design attributes (Table 4.6) using the Quality Function Deployment (QFD) house of quality matrix facilitates their presentation in design terms, and in a format that is independent of any design solution or materials specification. The use of a structured technique such as QFD, ensures that design attributes adequately reflect the wishes and priorities of the client. The presentation of requirements in a solution-neutral format also ensures that the same requirements set is available to the various disciplines involved in a project. It is therefore possible to adopt a CE approach to design.

4.6.4   Incorporation of CE principles

The methodology represented by the CRPM considers the life-cycle requirements of the facility (acquisition, operation and disposal information) early on in the process, by a multi-disciplinary requirements processing team. This exclusive focus on the wishes of the client ensures that the process is not overwhelmed by too much information, as would have been the case if other project requirements were included in CRP. The early consideration of life-cycle issues will also facilitate their incorporation in the design process. The key role and multi-disciplinary nature of the RPT also underscores the principle of CE with respect to collaborative working of multi-disciplinary teams.

4.7   Conclusions

The chapter has described a methodology (the CRPM) for incorporating the 'voice of the client' within a CE based design environment. It should be noted that the CRPM focuses only on client requirements (Table 4.1). The output of the CRPM (weighted design attributes that have been mapped to prioritised client requirements) in itself is not sufficient for design, but should be combined with other requirements (e.g. site, environmental and regulatory requirements) to provide a proper context for design. However, the focus on the client ensures that there is sufficient understanding of his/her needs to increase the likelihood that they will be balanced against the constraints (or possibilities) of other requirements.

Using the CRPM is also likely to involve more resources (time and people involved in the RPT) in the initial stages. But it is expected that the use of a multi-disciplinary team to precisely define the wishes and expectations the client, and the consideration of life-cycle issues up-front will provide savings in the longer term of the project and facility lifecycle.

4.8   References

CCC (2004), The Clients' Charter: http://www.clientsuccess.org.uk/ccc.asp, accessed 27th July 2004.

CIB (1997), Briefing the Team, Construction Industry Board (CIB), Thomas Telford, London.

CIT (1996), Benchmarking Best Practice Report: Briefing and Design, Construct IT (CIT) Center of Excellence, Salford, UK (ISBN: 1–900491–33–8).

Evbuomwan, N. F. O. and Anumba, J. C. (1995), 'Concurrent Life-Cycle Design and Construction' in Topping, B. H. V. (ed.), *Developments in Computer Aided Design and Modelling for Civil Engineering*, CIVIL-COMP PRESS, Edinburgh, UK, pp. 93–102.

Farbstein, J. (1993), 'The Impact of the Client Organisation on the Programming Process', in Preiser, W. F. E. (ed.), *Professional Practice in Facility Programming, Van Nostrand Reinhold*, New York, pp. 383–403.

Fellows, R., Langford, D., Newcombe, R. and Urry, S. (1983), *Construction Management in Practice*, Construction Press: London

Gibson, G. E. Jr, Kaczmarowski, J. H. and Lore, H. E. Jr, (1995), 'Pre-project Planning Process for Capital Facilities', *Journal of Construction Engineering and Management*, Vol. 121, No. 3, pp. 312–318.

Goodacre, P., Pain, J., Murray, J. and Noble, M. (1982), Research in Building Design, Occasional Paper No. 7, Dept. of Construction Management, University of Reading, UK.

Hannus, M. (1992), 'Information Models for Performance Driven Computer Integrated Construction', CIB (International Council for Building Research) Proceedings Publication 165, Vanier, D. J. and Thomas, J. R. (eds), National Research Council, Ottawa, pp. 258–270.

Highways Agency (2000), Route Management Strategy Guidance (available at: http://www.highways.gov.uk/info/rootstrat/rmsgm1/index.htm).

Howie, W. (1996), 'Controlling the Client', New Civil Engineer, 17 October, p. 12.

Integration Definition (IDEF) 1993, 'Integration Definition for Function Modelling (IDEF-0)', Federal Information Processing Standards Publication 183, National Institute of Standards and Technology, USA.

Kamara, J. M. and Anumba, C. J. (2000), 'Client Requirements Processing for Concurrent Life-Cycle Design and Construction', *Concurrent Engineering: Research & Applications*, Vol. 8, No. 2, pp. 74–88.

Kamara, J. M. and Anumba, C. J. (2001), 'A Critical Appraisal of the Briefing Process in Construction', *Journal of Construction Research*, Vol. 2. No. 1, pp. 13–24.

Kamara, J. M., Anumba, C. J. and Evbuomwan, N. F. O. (2000), 'Establishing and Processing Client Requirements – a Key Aspect of Concurrent Engineering in Construction', *Engineering, Construction and Architectural Management*, Vol. 7, No. 1, pp. 15–28.

Kamara, J. M., Anumba, C. J. and Evbuomwan, N. F. O. (2002), *Capturing Client Requirements in Construction Projects*, London: Thomas Telford.

Karhu, V., Keitilä, M. and Lahdenperä, P. (1997), 'Construction Process Model: Generic Present-State Systematisation by IDEF-0', VTT Tiedotteita-Meddenlanden-Research Notes 1845, VTT Building Technology, Espoo, Finland.

Kometa, S. T., Olomolaiye, P. O. and Cooper, P. A. (1995), 'Project Success and Pre-Contract Client Evaluation', in Thorpe, A. (ed.), Association of Researchers in Construction Management (ARCOM): Proceedings of the eleventh Annual Conference held from 18–20 September at the University of York (Vols 1 and 2), pp. 270–276.

Kuan, K.-K. (1995), 'Facilitating Conceptual Design in Concurrent Engineering', Proceedings of the CE 1995 Conference, McLean, Virginia, Kamara, J. M. and Anumba, C. J. (2001), 'A Critical Appraisal of the Briefing Process in Construction', *Journal of Construction Research*, Vol. 2, No. 1, pp. 13–24, 223–229.

Kumar, B. (1996), A Prototype Design Brief Development Assistant, MSc Dissertation, University of Glasgow, UK.

MacLeod, I. A., Kumar, B. and McCullough, J. (1998), Innovative Design in the Construction Industry, Paper 11180, Proceedings of the Institution of Civil Engineers (Civil Engineering), Vol. 126, pp. 31–38.

Mallon, J. C. and Mulligan, D. E. (1993), 'Quality Function Deployment – A System for Meeting Customers' Needs', *Journal of Construction Engineering and Management*, Vol. 119, No. 3, pp. 516–531.

Marks, R. J. Marks, R. J. E. and Jackson, R. E. (1985), *Aspects of Civil Engineering Contract Procedure (3rd Edition)*, Pergamon Press, UK.

Newman, R., Jenks, M., Dawson, S. and Bacon, V. (1981), Brief Formulation and the Design of Buildings: *A Report of a Pilot Study*, Buildings Research Team, Department of Architecture, Oxford Brookes University, UK.

Othman, A. A. E., Hassan, T. M. and Pasquire, C. L. (2004), 'Drivers for dynamic brief development in construction', *Engineering, Construction and Architectural Management*, Vol. 11, No. 4, pp. 248–258.

Palmer, M. A. (1981), *The Architect's Guide to Facility Programming*, America Institute of Architects, New York.

Parsloe, C. J. (1990), *A Design Briefing Manual, Application Guide* 1/90, Building Services Research Institute Association (BSRIA), UK (ISBN 0 86022 266 7).

Perkinson, G. M., Sanvido, V. E. and Grobler, F. (1994), 'A Facility Programming Information Framework', *Engineering, Construction and Architectural Management*, Vol. 1, No. 1, pp. 69–84.

Prasad, B. (1996). 'Concurrent Function Deployment – An Emerging Alternative to QFD: Conceptual Framework', in Sobolewski, M. and Fox, M. (eds), *Advances in Concurrent Engineering*: Proceedings of CE96 Conference, Technomic Publishing Company, USA, pp. 105–112.

Rezgui, Y., Bouchlaghem, N. and Austin, S. (2003), 'An IT based approach to managing the construction brief', *International Journal of IT in Architecture, Engineering and Construction*, Vol. 1, No. 1, pp. 25–28.

Salisbury, F. (1990), *Architect's Handbook for Client Briefing*, Butterworth Architecture, London.

Sanvido, V. E. and Norton, K. J. (1994), 'Integrated Design-Process Model', *Journal of Management in Engineering*, Vol. 10, No. 5, pp. 55–62.

Schenck, D. and Wilson, P. (1994), *Information Modelling the EXPRESS Way*, New York: Oxford University Press.

Serpell, A. and Wagner, R. (1997), 'Application of Quality Function Deployment (QFD) to the Determination of the Design Characteristics of Building Apartments', in Alarcón, L. (ed.), *Lean Construction*, A. A. Balkema, Rotterdam, pp. 355–363.

Strategic Forum for Construction (SFC) (2002), Accelerating Change, Strategic Forum for Construction, available from: http://www.strategicforum.org. uk/pdf/report_sept02.pdf, accessed 27th July 2004.

Vanier, D. J., Lacasse, M. A. and Parsons, A. (1996), 'Using Product Models to Represent User Requirements', Construction on the Information Highway: CIB (International Council for Building Research) Proceedings, Publication 198, 511–524.

Worthington, J. (1994), 'Effective Project Management Results from Establishing the Optimum Brief', Property Review, November, pp. 182–185.

Yusuf, F. (1997), 'Information and Process Modeling for Effective IT Implementation at the Briefing Stage', PhD Thesis, University of Salford, UK.

Chapter 5

Procurement, contracts and conditions of engagement within a Concurrent Engineering context

Peter Walker

5.1 Introduction – procurement systems

The approach to contractual arrangements for building in post-war Britain has been characterised by a period of stability, during which standard building and consultancy contracts have been built around established procedures and traditions. These standard contracts have been extensively documented in various texts (e.g. Chappell and Powell-Smith, 1997; Cox and Thompson, 1998; Murdoch and Hughes, 2000), are generally well understood by the participants, and importantly have been (regularly) tested in the courts. As documents they are now more than simply the expression of the legal intention of the parties; they have become (amongst other things) procedural manuals that provide an agenda for the actions of the various actors in a construction project.

Of the many standard forms of contract available, the most familiar is that published by the Joint Contracts Tribunal (JCT, 1998). Changes in modes of procurement have encouraged the JCT to publish contracts forms for Management Contracting and Design and Build contracting. Although the JCT forms dominate, they increasingly compete with other standard forms for example the New Engineering Contract (NEC, 2002) suite of contracts and PPC2000 (ACA, 2000), the so-called Partnering Contract. It is common for standard forms to be amended, in some cases to the point of being unrecognisable; and non-standard forms specifically drafted for a particular client or project are increasingly used.

Not withstanding the alternative forms and the wide use of amended wording, all construction contracts address similar matters and typically contain terms relating to:

- Details of the parties and description of the Works;
- The employer's representative (or contract administrator) and his powers of instruction;
- Obligations (of Contractor and Employer) and sanctions for non-fulfilment;

- Time, Payment and arrangements for claims for extra time and payment;
- Liabilities and Insurance;
- Quality of materials and workmanship (and, where appropriate Design);
- Health and Safety;
- Disputes and Termination.

Alongside this, the professional institutions (RIBA, RICS, ACE) have for many years produced model standard contracts for the engagement of consultants.[1] Traditionally the appointment of the consultants is made directly by the project promoter and the contract is directly between these two parties. The construction contactor, sub-contractors and suppliers are not parties to the contract. Forms of contract have been published recently that aim to integrate design and construction services more closely; for example the NEC suite of contracts now includes a form of consultant agreement (NEC, 2002) and PPC 2000 (ACE, 2000) allows for the creation of a project alliance with all parties in an integrated contractual nexus. These changes have, of course, arisen as a result of demand – in recognition of the new ways in which construction procurement is organised – but are not appropriate for traditional contracting. However, new, standard and bespoke forms all tend to follow a similar format. Typically the consultancy agreement covers the following matters:

- The parties;
- Consultant's Obligations – in which the consultant undertakes to discharge their duties using *reasonable skill and care*;[2]
- Fees – including periods for payment, variations, rights of set-off, deductions, additional payments;
- Intellectual property – copyright usually remains with the consultant who grants the client a license to use it for defined purposes;
- Insurance – an undertaking that the consultant has *and will maintain* a defined level of professional indemnity cover;
- Assignment – usually forbidding assignment on the part of the consultant and restricting the employers right;
- Dispute resolution – there is a statutory right to adjudication, as most agreements fall within the scope of the Housing Grants Construction and Regeneration Act, 1996;
- Deleterious materials – an undertaking not to specify or approve these;
- Suspension and determination – circumstances, procedures and payment;
- Jurisdiction – the 'nationality' of the courts and the legal system.

Other matters might include defining the key people to work on the project, confidentiality, special safety arrangements and an undertaking at some point in the future to enter into a collateral warranty. The scope of the service is normally set down in a separate schedule. This is of fundamental importance as this sets the project deliverables and is effectively what the consultant has priced for doing.

The procurement system can have important implications for the degree to which CE can take place or be encouraged to take place and the purpose of this chapter is to examine this thesis with particular regard to the appointment of the design consultants. The procurement system for the purpose of this chapter can be thought of as having two distinguishable elements: the relationships of the participants, and the contractual arrangements that bind all of them together. The two are of course inter-linked. It is also useful to make an occasional 'artificial' split between the way these elements relate to the project's consultancy services and to its physical construction. The two are of course inter-related and indeed it could be said that the recent trend in the UK construction industry is towards their ever more intimate linking. A useful way to classify procurement systems is as *separated* – in the sense that the design and construction functions are separated; *integrated* – where there is some varying degree of integration of the design and construction functions; and *management or mediated systems* – where a management function takes place in relation to the design and construction functions (see for example, Masterman, 1992 and Winch, 1996).

5.2   Separated systems

5.2.1   *Traditional general contracting*

In traditional procurement, the design and other consultancy services are provided by independent or in-house designers (Architect, Engineer) in direct contract with the building promoter, while a separate contract for the construction of the project is placed with a contractor. The consultant team see the project through the various stages from establishing the feasibility and financial viability of the project, through the design (including negotiating and securing statutory approvals); preparing the production information – the specifications and drawings for the building; assembling the tender and contract documents; and inspecting and supervising the works on site. The system is characterised by a complex web of contracts between the employer, his agents, designers and a main contractor, whose involvement and responsibility is limited to the construction phase of the project. The system is separate in at least two important respects. The organisations that carry out the work are *separate firms* and the temporary project contractual arrangements do not set out to address this. There is a *separation of time*; the design consultants are appointed some time in advance of the constructors – indeed the identity of the construction contractor is generally unknown and their appointment will usually only follow on from a competitive tendering process.

For the client, the advantages of this system lie in the fact that design is retained by its appointed designers, affording the possibility of close control both of the specification and of the realisation of the product. Advantages to the supply-side (main-, sub-contractors and suppliers) include limited liability and relatively low transaction costs (in particular the reduced cost of bidding against a set of clearly defined and settled specifications). The system is tried and tested and well understood by the parties.

The perceived disadvantages lie in the separation of the design, procurement and construction phases. This adds time to the process. It also limits opportunities for a concurrent approach in that early and direct communication between designers, contractors and (in particular) their specialist suppliers is difficult although not impossible – 'nomination' and the use of the Contractors Design Portion supplement (JCT, 1998), see later, allow some degree of designer/constructor integration. The project may therefore miss out on the benefit of the specialised knowledge of constructors and manufacturers, particularly concerning buildability and value for money. The arrangement is normally combined with lump-sum reimbursement and selective tendering to form the procurement system described in the UK as 'traditional contracting'.

5.2.2   The consultant appointment in traditional contracting

The appointment of the consultants is made directly by the project promoter and the contract is directly between these two parties. The scope of the consultant's service can be considered in three areas (which are also three sequential stages of the process) – producing from the brief the concept design; producing from the design the production drawings and specifications; and preparing the tender documentation, managing the tender process and administering the construction contract. The design work is completed prior to tender (in practice usually substantially rather than absolutely complete). Standard terms of appointment do not place any explicit duty on the designer to consult with the contractor or suppliers. In practice there will typically be some consultation with suppliers of materials and components as part of the process of preparing the design and specifications.

The production information drawings and specifications are not only descriptors of what is to be built, but are also contractual documents (incorporated by reference). Any changes introduced after the contract has been entered into, either by the designer or by the contractor, for example to increase buildability or to reduce cost, will require the issue of a contractual variation order. Changes by the contractor are in practice unlikely. Some of the reasons for this are as the following.

As the contractor and the construction supply chain arrive late in the process they can only make recommendations for change when the design is fully developed. Such changes introduced at this stage are likely to be of limited benefit particularly when any design costs and any disruption to the programme is taken into account (although there are exceptions to this).

Under typical traditional contractual arrangements the contractor has no duty to propose changes to the design – indeed the contractor has no duty even to check for errors. UK contractors generally do not have design departments or design skills within the organisation and therefore are not set up in such a way as to review, audit or question the design. As the contractor is not party to the pre construction decision making process he will have little or no knowledge of designers brief or the rationale for the design.

Under normal traditional contractual arrangements the contractor has no commercial incentive to propose changes to the design. Whilst changes to the design might make the building cheaper to construct, the contract terms generally provide that the financial benefits will come to the promoter rather than the contractor thus removing any commercial motivation.

5.2.3   Opportunities for a CE approach in traditional procurement

As traditional procurement is predicated on a sequential approach with a clean contractual break between design and building, it would be surprising to find any aspects of this that allow or create opportunities to adopt a CE approach. In a desperate search in such an inhospitable environment is it possible to turn up any ways in which to a degree or in part traditional contracting allows for or encourages CE?

The only two possible ways in which there may be some degree of concurrency in traditional design are in the use of the process of *nomination*[3] and in the use of the *Contractors Designed Portion* supplement – but do these amount to a CE approach or even a partial concurrent approach? Kamara (2003) helpfully sets out the 'two key principles [of concurrent engineering] as 'integration and concurrency'. Integration here is in relation to the process and content of information and knowledge, between and within project stages ... [and] ... also involves upfront requirement analysis by multidisciplinary teams and early consideration of all lifecycle issues affecting a product. Concurrency is determined by the way tasks are scheduled and the interactions between different actors (people and tools) in the product development process.' (Kamara, 2003).

Nomination involves the early appointment of a specialist sub-contractor or supplier *prior to the appointment or even identification of the main contractor*. The main contractor is directed (by the designer – the employers agent) to enter into a contract with the nominated sub-contractor. Nomination exists within traditional contracts to allow the upstream involvement of specialist suppliers and sub contractors (but significantly not the main contractor) in the design process; in this respect it can be said to schedule the design task in a way that allows the interaction of some other actors in the product development process. However the absence in this arrangement of the principle construction actor – the main contractor, responsible for the integration and management of the building process – creates a fundamental barrier to true concurrency. The process can be said to involve the upfront analysis by a multidisciplinary[4] team although this is generally *realisation analysis* of a part of the product rather than *requirement analysis*; the overall requirement having been established even earlier in the design process. Although this allows some integration and concurrency, it cannot be said to truly be a CE approach.

The use of the Contractors Design Portion Supplement (CDP) used with JCT standard forms of contract allows a part of the works to be designed by the contractor, or more usually, by a specialist sub-contractor. Importantly it creates within the building contract a design contract, with the building contractor taking on the same skill and care responsibility and liability as a designer for the element of the works covered by the CDP. The CDP is properly used where a part of the works is the subject of a performance specification. In fact it satisfies neither the concurrency test – there is no concurrency in the design process, the design is simply a response to the performance specification; or the integration test – the design is not the work of a multi disciplinary team working together. Indeed in artificially isolating one aspect of the building design from the others it could be said to be positively the antithesis of CE.

Separated systems have many positive qualities – Winch suggests that it offers cost certainty (after tender), good quality assurance and transparency in the formation of the contract (Winch, 1996). Done properly, it does achieve all these- what it does not do is to facilitate a CE approach to design. In a later paper on a similar theme Winch highlights the problem of over engineering in separated systems where the consultants, to protect there own position, are 'obliged to specify the product completely'. Yet their '[inevitable] lack of experience with site processes means that their specification decisions do not reflect site conditions or the capabilities of contractors' (Winch, 2000). The commercial contractual obligation of the designer promotes an approach that discourages good design.

5.3   Management systems

5.3.1   Management contracting, construction management and design and management

The concept of procuring a project's management input separately is based on the fact that most of the construction and indeed much of the design is procurable from specialist 'works' or 'trade' contractors, leaving the traditional main contractor free to engage in a consultant-managerial role. Management Contracting (MC) is often preferred where time is of major importance or the project is complex. Construction Management (CM) is a variant that originated in the USA and is common in many other countries. Its most significant feature is the direct contractual links between the Client and the 'trade contractors'.

The main advantage of a management procurement system is speed. There are two reasons for this. First, the relatively 'open' nature of its commercial relationship with the client means that the management organisation (whether MC or CM) needs little lead-in time as a prelude to its involvement in the project and can thus work at an early stage with client consultants. Second, the fact that construction is carried out in specialist works or trade packages (overseen by the management organisation) means that their design, procurement and construction periods can be overlapped. As with the traditional procurement approach, the client retains control of the design team, thus avoiding compromise on quality, while maintaining flexibility in terms of managing and incorporating change.

The management organisation itself carries little or no financial risk for the project, and consequently can assume a more independent and 'professional' role in the project. On the other hand, in business there is normally an association between risk and reward: the management role attracts relatively low fees in line with its low associated risks. The main disadvantage for the client is in regard to cost certainty. Management systems have historically been associated with the so called 'cost-plus methods of reimbursement', and choosing a conventional management system has been considered as precluding the client from the comfort of a either a lump sum or guaranteed price. The same openness, speed and flexibility that characterises the relationship brings with it a lack of financial certainty prior to the commitment to build, and a lack of 'single point responsibility'.

5.3.2   The consultant appointment in management or mediated procurement

The promoter would usually appoint the design consultant on terms similar to those in traditional contracting. The scope of consultancy service would vary in that the design output would be presented in such a way as to allow the works to be tendered and constructed by specialists in separate trade or works packages. However this is generally simply a matter of editing the assembly of the information, rather than a matter of substance. More significantly, the designers would work alongside the MC or CM in developing the design and would take their input into account in developing the designs and specifications. The MC or CM would in theory be appointed at the outset. In practice it is often the case that the decision to adopt this form of procurement is not made until relatively late in the design sequence and therefore the construction appointment often comes late in the project design development process.

5.3.3 *Opportunities for a CE approach in management or mediated procurement*

There are undoubtedly greater opportunities for an integrated approach, however the input of the management contractor is most often in the areas of design *administration* (controlling and distributing information) and in what could be called *brokerage and commercial management* (identifying, negotiating and managing suppliers and sub-contractors) rather than in developing the design solution. Whilst this will have an indirect influence on the design, this will often be restricted to issues of individual material, component or assembly selection and will have little fundamental impact on for example the form or planning of the building. In practice this influence is often exerted to reduce the project costs rather than to increase value for the promoter (although it could of course be said that reducing the cost is a form of added value, but only where the savings are passed on to the promoter).

In this sense MC and CM could be said to allow a degree of integration in relation to the process and content of information and knowledge within project stages, but probably only to a limited degree between project stages (in the sense of this being restricted to the latter stages of design). It would generally not involve upfront requirement analysis by a multidisciplinary team in that requirement analysis would generally not fall within the remit or skill of the MC or CM. This is also true for the early consideration of lifecycle issues affecting a 'product'.

The organisation of mediated or management procurements systems in the UK generally precludes the meaningful involvement of the contractor in the development of the design or requirement analysis due to the timing of their involvement, the limited design skills base within the organisation and the management functional remit they are given.

5.4   Integrated systems

5.4.1   *Design and build*

Design and build became increasingly popular in the UK during the 1980s and 90s (Franks, 1984; Gray *et al.*, 1994). The contracting organisation undertakes both the design and construction of a project. There are many variations in the way in which consultancy services are delivered: there being various possible levels of employer-involvement, ranging from 'pure' design and build (where the contractor has the opportunity to control the substantial part of the project's design) to extreme situations where the entire design control is retained by the employer, but the ultimate liability for the design is passed to a contractor. It is frequently said, particularly by design and build contractors, that this arrangement is not driven by a desire to increase quality or encourage innovation, but by a wish on the part of promoters to pass on risk. Package deals or turnkey projects and integrated design and build – where the design and build functions are carried out within the same firm – are rare in the UK.

A number of advantages are claimed for integrated systems. For the client, there will be less upfront expenditure on contract documentation and the likelihood of a more economic design for construction (though savings may not necessarily be passed on to the client). There will be single point responsibility, with the possibility of passing all major risks (including those of maintaining cost and time certainty) to the supply side. Perhaps most importantly, the integration of design and construction within the same team permits much shorter overall project time periods. Despite the disadvantages of higher bidding costs and risks, the system offers the contractor almost total control of all aspects from design to commissioning. This offers scope for 'value engineering' and 'buildability', and an enhanced opportunity to manage risk in return for reward.

The disadvantages to the client are in the loss of direct control at the point at which design responsibility is transferred to the contractor. The contractor's commercial objective will invariably be economy of design, though appropriate output specifications, in the form of a considered and effective set of Employer's Requirements should guard against quality being compromised. The system also demands earlier and firmer commitment to these requirements. The client may not enjoy the luxury of flexibility since giving up control of the design may mean that any post-award changes in client's requirements will be difficult, or at least relatively expensive.

5.4.1.1   The consultant appointment in design and build contracting

The most popular way of appointing consultants for design and build projects in the UK involves the novation – or switch – of the design team from the building promoter to the building contractor (Chappell and Powell-Smith, 1997). Typically the design team prepare the scheme design – general arrangement plans, elevations, sections and outline or performance specifications – on behalf of the building promoter. These form the basis of the tender documentation (referred to in standard JCT forms as the Employer's Requirements) on which the tendering contractors base their offer (the Contractor's Proposal). Once the contract is let the design team (but normally not the quantity surveyor) is novated to the building contractor. The design team is then retained by the contractor until the conclusion of the project, and produces the production information on the instructions of the contractor. The service delivered by the design team for the contractor differs fundamentally from that which would have been delivered for the employer in that the function of the drawings and the specifications is simply to describe what is to be built. In contrast to *separated systems* of procurement the design drawings and specifications produced for the contractor are not contract documents (in the sense of the building contract). For this reason the drawings are always less detailed and therefore cheaper for the consultant team to produce.

The appointment terms of consultants for a design and build contract are much the same as for traditional procurement, although the scope of work generally excludes any involvement in administering the construction contract. Where it is intended that the responsibility for the project's consultants will switch from the employer to the contractor there will also usually be a Novation Agreement. This is a relatively straightforward document that simply sets out the terms of such a switch.

5.4.1.2   Opportunities for a CE approach in design and build procurement

On the face of it the appointment of the consultant initially to prepare the outline design for the employer – the actor best placed to define their product requirement; followed by appointment by the contractor – the actor best able to decide how the product is to be delivered, is a sensible and logical way to proceed. In reality the early outline design casts the die in such a rigid way as to make latter changes to the detailed design difficult and costly to effect. Complications arise where changes to the design require third party approvals (e.g. changes that would require a new or amended planning permission as required under town and country planning legislation); or where changes would affect the terms of the contract between the contractor and employer (changes to the employers requirement). The terms of appointment and the fees of consultants are settled with the employer prior to novation and there is no commercial incentive for the design consultant to undertake further reviews of the basic design with the contractor.

5.4.2   Partnering and framework agreements

The conventional view that construction projects are sets of one-off, discrete, contractual deals is not always the correct one. In the UK, writers like Cox and Thompson (1998) have described the trend to longer-term arrangements such as framework agreements and 'serial' or 'strategic' partnering which involve long-term relationships over programmes of work rather than an individual project.[5] The advantages lie in saving the costs of re-bidding each individual project, the prospects of continuous improvement from one project to the next, and a more predictable workflow for the supply-side. Disadvantages include the chance of relationships becoming too comfortable, and the client's loss of access to 'market value' that comes with abandoning repetitive tendering. To offset these, such deals often include incentives or performance improvement regimes. In some of these longer-term agreements[6] a competitive element is retained. Indeed legislation may require this – publicly funded works in the UK must be advertised and competed for in accordance with the UK Public Procurement Regulations, as required by European Community rules.

Whether relationships are extended or one-off, they need to formed in the first place, and this is accomplished with a greater emphasis on either competition or co-operation. At the competitive end of the spectrum open and selective tendering rely on price competition as their main or only criterion. However, some clients adopt a more co-operative outlook and favour negotiation, where non-price criteria play a significant part. Many decision-makers use 'weighted scorecards' to achieve the balance between price and quality that is appropriate for their project. Paradoxically, as client-contractor relationships have become closer, the client-consultant relationship has become more competitive. The traditions of mandatory fees and standing relationships have given way to competition and formality. The advantages and disadvantages of the various approaches are fairly self-evident: it is often argued that 'lowest tender does not equal best value'. Since 1999 this philosophy has dominated public-sector procurement in the UK. On the other hand the benefits of 'taking the lowest bid' can be very persuasive, especially in terms of demonstrating financial probity. This philosophy is at the heart of European public procurement rules.

Project-specific partnering[7] as the term suggests, relates only to a project, and is based on a change of attitudes of the participants, sometimes involving 'open-book' costing. However, it is misleading to assume that a more co-operative procurement has taken place as the co-operation may be entirely post-award and follow an intensely competitive tendering process.

5.4.2.1   The consultant appointment in a partnering arrangement

Until the arrival of PPC2000, partnering projects were carried out using existing standard contracts, for example the JCT With Contractors Design form of Contract (JCT). In these cases the consultants would be appointed using either the industry standard forms or bespoke forms of appointment. The terms of the consultant appointment would not need to be different in any significant way as the services would be consistent with those required for other forms of procurement. There would in some cases be a contractual requirement to participate in value management workshops and there would be the need to integrate the work of a wider group of participants in the design process. The partnering charter would deal with what could be described as 'behavioural issues' in terms of the relationship between the various participants. Quite what legal status this charter has is a matter of debate. In some partnering arrangements the fees of the consultant would be based on reimbursable rates with payment based on the hours spent reclaimed at regular intervals throughout the contract. The contract would normally contain a detailed schedule of what constituted the input costs of the consultant and the procedures for applying a profit mark up to these. There would often be a mechanism for shared rewards (typically savings made on the construction contract costs) and in the case of a strategic partnering arrangement, a method of discounting fees in relation to the volume of work provided.

PPC2000 (ACA, 2000) provides an integrated form of consultant appointment (that is to say integrated with the contractor and others). It also introduces a number of other partnering procurement concepts: for example an obligation to work together and individually in the interest of the project, and an obligation to work in a spirit of trust, fairness and mutual co-operation. The client is treated as an active project team member with defined duties and obligation; and other actors not usually parties to construction contracts, described as 'interested parties' are included. These include for example the local authority and the body providing funding for the project. There is reference to a 'Core Group' consisting of the client, constructor and designer and to a 'Partnering Adviser', responsible for facilitating the partnering approach where this breaks down. Value management, incentives and the pricing framework are all defined, and the partnering timetable and the project timetable are both made contract documents. The contract recognises and allows for the integration of the supply chain, and is founded on the idea of the early appointment of both the construction and design team. The client is also able to propose changes during the construction stage.

5.4.2.2 Opportunities for a concurrent approach in partnering procurement

Partnering, as outlined earlier creates the project environment and provides the opportunity by its contractual organisational arrangements, for integrated concurrent design. How does partnering perform when tested against the two key CE principles of integration and concurrency? It can be said that partnering allows the integration both of the process and the content of information and knowledge. It can also be seen that by the involvement of the total project team over the life of the project, that this be will possible between and within project stages. The early involvement of the designer, contractor and others will also allow, at least the possibility of early consideration of all lifecycle issues affecting the project product, and at least go some way to meeting the concurrency test.

5.5   CE and commercial constraints

There are a number of ways in which construction consultants may be reimbursed for their services. Some of these will create a greater commercial incentive to expend the time required to consider all the lifecycle issues affecting the project and to spend the time developing a concurrent approach to design.

Reimbursable fees (payment based on the hours worked multiplied by pre agreed hourly rates) are common in many professions but have been relatively unusual in construction up until the advent of open-book partnering arrangements, where the consultant's fee may be arrived at by adding an agreed overhead recovery and profit mark-up to the basic hourly cost (i.e. the salary and direct on-costs of individual staff). This payment regime encourages the time to thoroughly review all aspects of the project from the consultants commercial perspective, but may place the overall project budget at risk of overrun or lead to an actual overspend.

Percentage fees are based on a percentage multiplier applied to the construction cost, thus presuming a direct relationship between this and design complexity. It is a method that has clear shortcomings for both parties: for the designer, it fails to recognise the time and ingenuity involved in producing an economic design; and to the client, it appears to reward the consultant for an extravagant design. It would also create a commercial disincentive to, for example, spending time reviewing all the life cycle issues affecting the project product.

Lump sum fees, possibly fixed to be effectively a 'guaranteed maximum price', will potentially limit the designer's efforts when they have reached a certain level.

In some cases the form of reimbursement may be mixed – for example the early feasibility study (which may include consideration of all the lifecycle issues affecting the project) – may be paid for on a reimbursable basis and converted to a percentage fee or lump sum once the brief for the building is finalised.

Less common methods of payment include contingency fees, where payment is made dependent on some condition or performance: for example, a developer may agree to pay the consultant fees only if planning permission or land purchase is secured. Fees based on value added however, are extremely rare in the construction industry although it is easy to imagine how these could be arrived at. For example, a commercial property developer may pay an enhanced fee for a building designed in such a way that it achieves a more dense site coverage, or for a building that lets quicker or on better commercial terms.

Occasionally consultants may be employed on a regular retainer – often paid on a monthly basis. This would be appropriate where a consultancy service is provided on an intermittent but regular basis, for example an architect providing a consultant architect service to a local planning authority. Payment of royalties is common in the private housing sector where a house builder pays architects for the use of a standard house design when it is used.

What all these fee arrangements share is payment based on the input value, that is, the time spent on doing the task, rather than payment based on the value added by the service or the value added to the finished product. It is possible to speculate that commonly used fee payment arrangements for design consultants based on input costs rather than output added value are unlikely to encourage a CE approach. It is also possible to suggest that it is likely that these arrangements represent a significant disincentive to a more considered and integrated approach to the organisation of the project design.

5.6   Reflections on partnering and CE

The choice of procurement system and contractual arrangements for both designers and builder can have a profound effect on the ability to adopt a CE approach. Of the three systems examined, integrated systems such as design and build or partnering are on the face of it most likely to encourage a CE approach. However on closer examination, the popularity of the consultant switch or novation process in design and build procurement in the UK, has the effect of dis-integrating the design and construction process. This leaves partnering or framework agreements as the procurement systems most conducive to a CE approach. But this is not explicitly the benefit sought in partnering.

Long term strategic partnering arrangements offer a number of benefits for the consultants within this arrangement. These could arise from the volume of work, the regular flow of work or increased efficiency arising from a better understanding of the clients business and their building needs. Construction is generally a project based cyclical industry: the balancing of capital and resources to cope with unpredictable fluctuations in demand can create real business difficulties for consultants. Against this backdrop the attraction of a longer-term agreement, with regular payments and a greater degree of predictability and certainty in workflow, can be clearly seen.

The benefits to the consultants are clear, but why do clients enter into these longer-term agreements? The process of placing advertisements, sifting through the applicants, short listing, interviewing and eventually agreeing terms with the successful firms is time consuming and costly;

so why bother? The answer, as with almost all business processes, is that the return justifies the investment. The investment is easy to identify (and indeed relatively easy to quantify) it is the management time and expense that the promoter goes to in establishing the preferred contractor; but the identifying and quantifying the return requires a little more effort. It is of course the case that for different promoters in different businesses the benefits will vary, but many will consistently occur. Some of these are:

Quality improvements   This has two distinct components. First, a better quality product: buildings that perform better as a result of the better understanding of the promoter's business needs, aspirations and functional requirements. And secondly, a better quality process: a way of working together that adds value by eliminating the problems that arise in temporary project teams. These problems include a lack of common management procedures and compatible systems; a lack of inter personal understanding (usually accompanied by a high degree of inter personal conflict!); a lack of common goals and culture or at least a lack of understanding of what these might be.

Reduced transaction costs   It could be argued that the initial costs involved in setting up the agreement are higher than those incurred for a single project. However, this is a one off cost with a payback that improves with each new commission placed during the life of the agreement. The total costs are therefore considerably lower. Savings arise in many ways: savings in time arising from a reduced mobilisation time, savings in fees, for example lawyers fees in preparing and agreeing forms of appointment for each new project; savings in management time, for example in the briefing and induction of new consultants.

Value improvements   The ability to expand learning from one project to another will increase the value delivered in the end product. The greater the global knowledge of the consultant team the greater their ability to eliminate non-value adding processes. This can manifest itself in many ways. For example, if the partnering architects understand the local authority planning constraints on the promoter's land, they will avoid exploring options that are unlikely to obtain planning approval. Similarly, if the partnering engineers have developed a good understanding of the ground conditions they can very quickly arrive at a realistic foundation design without the need to embark on very detailed investigations too early in the project.

Project cost savings   A better understanding of the client's budget will allow the design and cost consultant to develop solutions that spend the money wisely. For example, the client may have modest funds for the capital cost of building but may have access to grants and other income to support the running costs. This can be an important informer of the design and budgetary strategy.

A greater motivation to success   It is inevitably true that companies, like individuals, will invest more personal capital in relationships that have a long-term future. Companies will have a strong commercial incentive to work to the success of such an agreement to ensure that they get a good volume of work. Also relevant will be a) the fact that the consultant has invested in resources and established these at a level that recognises the partnering or framework workload, b) the saving in marketing and bidding costs that the consultants makes from repeat business rather than pursuing new contracts and c) the ultimate incentive of renewing or maintain a place on the consultant panel. The consultant will also inevitably be closely identified with the partnering client by the public and by peers and therefore the kudos and public relations value of successful projects will be high.

Partnering, framework or consultant panel agreements may be attractive in providing security and greater certainty but they do not represent a panacea to the problems that beset consultants in the construction industry. The difficulty of managing the upstream design process, of dealing with late changes to the client requirements and the inherent difficulty of each new development being effectively a prototype remain. However the ability to better forward plan the business cost and resources, potentially reduces business management time and allows more design and project management time – time that potentially adds more value to the finished product. Concurrency is determined by the way tasks are scheduled and the inter-actions between different actors (people and tools) in the product development process (Kamara, 2003). Many of the advantages of partnering are synonymous with and consistent with the benefits derived from a CE approach.

5.7 Conclusions

Research in CE has focused on both the tools and the environment for its implementation in the construction industry. It is possible to speculate that we are seeing the beginnings of fundamental structural changes in the UK construction industry environment; many initiated in response to Constructing the Team (Latham, 1994) and Rethinking Construction (Egan, 1998). It is also possible to speculate that these changes, which have impacted on organisational, cultural and commercial aspects of the construction industry have created a climate in which the concepts of CE may now move from a largely theoretical tool, to an effective and widely used applied management technique. It has been said that there are many generic tools which may not have the CE label but nonetheless, reflect concurrent engineering principles (Kamara and Anumba, 2002). Many of these can be seen in the contractual arrangements and organisational structures used in partnering projects. The separation of design and construction has long been presented as the root problem of construction, and has been explored by many (see for example, Ballard and Koskela, 1998). Traditionally the model for design in the UK construction industry has been sequential rather than concurrent and the organisation of architecture, construction and engineering firms has been fragmented rather than integrated. Whilst changes to this order can be perceived, sustaining this will require careful study and review of the commercial and contractual arrangements for design consultants as a part of the wider procurement system.

5.8   Notes

1 A current trend appears to be for clients to prefer their own bespoke form of agreement; it is interesting to speculate on the reason for this. On the face of it is understandable that clients may be sceptical regarding the equitable nature of agreements prepared by a particular professional body for its members. The RIBA Standard Form of Agreement published in 1992 came in for considerable criticism due to a perceived bias in favour of the architect. That it was little used is perhaps no surprise.
2 The normal standard for professional services (as distinct from the building contract standard of *fitness for purpose*).
3 The process in which the design team selects and contractually nominates a specialist sub-contractor or supplier for an element of the works.
4 Multi disciplinary in the sense of a team consisting of *more than one discipline*, not in the sense of involving *all disciplines*.
5 As opposed to 'project specific partnering'.
6 As in the BAA 'consultant panel' systems for example.
7 As opposed to 'serial partnering' (see earlier).

5.9   References

ACA (2000), Project Partnering Contract: the ACA Standard Form of Contract for Project Partnering, Association of Consultant Architects (ACA), London.

Ballard, G. and Koskela, L. (1998), 'On the agenda of design management research', *6th Annual Conference of the International Group for Lean Construction*, Guaruja, Sao Paulo, Brazil, 13–15 August (available at: http://www.ce. berkeley.edu/~tommelein/IGLC-6/index.html, January 2005).

Chappell, D. and Powell-Smith, V. (1997), *The JCT Design and Build Contract, 2nd ed.*, Blackwell Scientific Publications.

Cox, A. and Thompson, I. (1998), *Contracting for Business Success*, Thomas Telford Publishing, London.

Eccles, R. G. (1981), 'The quasifirm in the construction industry', *Journal of Economic Behavior and Organization*, 2, 335–357.

Egan, J. (1998), *Rethinking Construction*, Report of the Construction Task Force to the Deputy Prime Minister, HMSO, London.

Franks, J. (1984), *Building Procurement Systems*, The Chartered Institute of Building, Ascot.

Gray, C., Hughes, W. and Bennett, J. (1994), *The Successful Management of Design – A Handbook of Building Design Management*, Centre for Strategic Studies in Construction, University of Reading.

JCT (1998), *The Standard Form of Building Contract*, RIBA Publications, London.

Kamara, J. M. (2003), 'Enablers for Concurrent Engineering in construction', in Martinez, J. C. and Formoso, C. T. (eds), *Proceedings of the 11th Annual Conference of the International Group for Lean Construction*, 22–24 July, Blacksburg, Virginia, VA, pp. 171–183.

Kamara, J. M. and Anumba, C. J. (2002), 'Collaborative systems and CE implementation in construction', in Tommelein, I. D. (ed.), *Proceedings of the 3rd International Conference on Concurrent Engineering in Construction*, Berkeley, CA, 1–2 July, pp. 87–98.

Latham, M. (1994), *Constructing the Team – Final Report of the Government and Industry Review of Procurement and Contractual Arrangements in the UK Construction Industry*, HMSO, London.

Masterman, J. (1992), *An Introduction to Building Procurement Systems*, E&F.N. Spon, London.

Murdoch, J. and Hughes, W. P. (2000), *Construction Contracts: Law and management*, 3rd Edition, E&F.N. Spon, London.

NEC (2002), *The Engineering and Construction Contract*, Thomas Telford Services Ltd, London.

Williamson, O. E. (1975), *Markets and Hierarchies: Analysis and Antitrust Implications – A Study in the Economics of Internal Organization*, The Free Press, New York.

Winch, G. (1996), The Contracting System in British Construction: The Rigidities of Flexibility, Working Paper No. 6, Le Groupe Bagnolet.

Winch, G. (2000), 'Institutional reform in British construction: partnering and private finance', *Building Research and Information*, 28, 141–155.

Chapter 6

Process management for concurrent life cycle design and construction

Michail Kagioglou, Ghassan Aouad, Song Wu,
Angela Lee, Andrew Fleming and Rachel Cooper

6.1 Introduction

An overview of the academic literature in processes is enough to appreciate that there exist a number of different terms and concepts within this 'process' context. Furthermore, processes exist within organisations at different levels of detail and use, and expressed using different terminology. In addition, the nature of the process used often depends on the modelling technique used to design and document the process.

There is also a difference between processes as expressed above and process maps/models. Rosenau (1996) suggests that process models are an effective way to show how a process works and he suggests as a definition that:

> A process map consists of an X and Y axis, which show process sequence (or time) and process participants respectively. The horizontal X axis illustrates time in process and the individual process activities or gates. The Y axis shows the departments or functions participating in the process...

Beyond this convention there is little formality in the method used to represent a process. It could be argued that this definition applies to a number of process maps and 'procedures' and sometimes the X and Y axis functionality is reversed. However, this definition does not identify the number or nature of the different levels of detail in the process. Furthermore, since time is included as one of the process map axes, Rosenau (1996) assumes that all process maps have a definite beginning and end, where in reality some processes are continuous and the factor of time is 'less' important that the actual content of the process. As Ould (1995) suggests there are two types of processes:

1   the sort that start when necessary and finish some time in the future, and
2   the sort that is constantly running.

This chapter provides a background on the subject of processes, providing some definitions, explanations and interrelationships between varying processes within an organisation. In addition the chapter illustrates the needs for and the effects on organisations and how processes can enable the Lifecycle design and construction process (or product development) and how they should be managed. By product development processes we consider those that cover the whole lifecycle of a product/service, from conception of the idea to final operation and use. An example of such a process and associated enabling IT is described.

6.2   The process defined'

The term 'process' can have different meanings for different people depending on the sector, function and market within which they are operating. Talwar (1993) defines process as a 'sequence of pre-defined activities executed to achieve a pre-specified type or range of outcomes'. Harrington (1991) refers to a process as 'any activity or group of activities that takes an input, adds value to it and provides output to an internal or external customer. Processes use an organisation's resources to provide definitive results'.

Oakland (1995) states that 'a process is the transformation of a set of inputs, which can include actions, methods and operations, into outputs that satisfy customer needs and expectations, in the form of products, information, services or – generally – results'. Davenport (1993) states that 'a process is simply a structured, measured sets of activities designed to produce a specified output for a particular customer or market' and he continues stating that 'processes are the structure by which an organisation does what is necessary to produce value for its customers'.

Consequently, an important measure of a process is customer satisfaction with the output of the process. Zairi (1997) states that 'a process is an approach for converting inputs into outputs. It is the way in which all the resources of an organisation are used in a reliable, repeatable and consistent way to achieve its goals'.

Furthermore, Bulletpoint (1996) suggests that regardless of the definition of the term process there are certain characteristics that this process should have, including:

1   Predictable and definable inputs
2   A linear, logical sequence of flow
3   A set of clearly definable tasks or activities
4   A predictable and desired outcome or result.

However, the above is not normally the case in construction, where unpredictable and speculative works lack certainty and although the outputs are normally broadly defined, the final outcome is the result of an evolving process.

6.2.1   Process levels and types

In considering the processes within organisations it is important to distinguish between the various types and levels of processes that exist. This enables the effective communication and classification of processes whilst ensuring that all levels and types are 'striving' towards satisfying customers and therefore ensuring the organisation's survival and continuous prosperity. Peppard and Rowland (1995) identify three distinct types of processes within organisations: strategic, operational and enabling. Their operation and interrelationships are illustrated in Figure 6.1.

Peppard and Rowland (1995) further define those types as follows:

- *Strategic*   those processes that the organisation plans for and develops its future.
- *Operational*   those processes by which the organisation carries out its regular day-to-day functions.
- *Enabling*   those processes which enable the strategic and operational processes to be carried out.

The latter definitions further illustrate the interrelationship of the three different types of processes. For example, if the goal of an organisation is to satisfy customer requirements by developing new products then a new product development (NPD) process needs to be in place at a strategic level. The operational processes will belong within this NPD process and enabling processes such as communications and IT will enable the effective implementation of the operation and strategic processes. Therefore, it can be seen that although the formulation of those processes is performed based on a top down approach, that is, strategic, operational, enabling, their implementation is bottom up that is, enabling, operational and strategic.

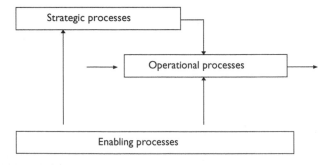

Figure 6.1 Types of processes.

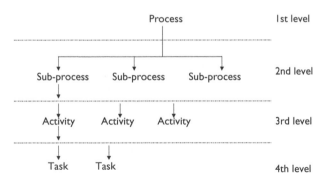

Figure 6.2 Process levels.

The top down approach to planning processes and the bottom up implementation can be further examined by considering the different levels in which processes 'exist', that is, process decomposition, as shown in Figure 6.2.

The different levels of processes can be further illustrated when considering the development of new products as an example. The 1st level in developing a new product development process will be to consider the NPD process as a whole from the conception of the idea or the identification of the need until the final commercialisation and withdrawal of the product. The 2nd level will consist of the different sub-processes such as design, engineering, manufacturing, marketing and other sub-processes. The 3rd level will consist of the different activities within those processes such as the design of a product, the design of the manufacturing process etc. The 4th and final level will consist of the different tasks that need to be performed to undertake the activities such as simulation, market survey, etc.

6.2.2 Process management

Organisations that wish to undertake improvements in productivity, quality and operations need to consider the management of these processes that will bring about these improvements (Elzinga *et al.*, 1995). A number of initiatives have appeared throughout the last three decades, which aim to define design and optimise these processes. Lee and Dale (1998) have identified this process orientation in the European Foundation for Quality Management (EFQM) and Malcolm Baldrige National Quality Award (MBNQA) models for business excellence and performance. Furthermore the study of processes has been expressed in different terms like, 'process simplification', 'process improvement', 'process re-engineering' and 'process redesign' (Lee and Dale, 1998). All the latter movements in process improvement can be defined in two main streams:

1   Management and continuous improvement of existing processes
2   Designing or redesigning of new processes.

The first stream has been examined in detail within the total quality management (TQM) literature and the second stream is mainly considered under the heading of business process re-engineering. The first stream aims to optimise and continuously improve, on-going, processes that have been in operation within an organisation, whereas the second stream aims to redesign/re-engineer processes, which either are not performing very well or which have been overtaken by the market forces. An example illustrating the need for both approaches can be seen, from a product viewpoint, in the computer industry. For example, in the early 1990s a '486' computer was considered 'state of the art'. As time and technology progressed new components were added to those computers to optimise their performance. However, the introduction of the Pentium processor meant that the '486' computer was obsolete and the total replacement of the computer was required. In the same way some processes are continuously improved but increased competition, technology improvements, etc. means that those processes need to 're-engineer' themselves to accommodate those changes. Furthermore, regardless of continuous improvement or re-engineering a certain change in the organisation is occurring.

6.3   The design and construction process

Construction is a project-based industry and as such it generates value by developing new products all the time either in terms of new buildings (both greenfield and brownfield) or by improving the built environment where the society members interact. The large number of actors that take part in every project often promotes a sense of 'adhocracy' whereby professional specialisms take charge of large parts of projects and the skills of a project manager are used to bring diverse groups of people together to deliver complex and demanding processes and structures (Anumba *et al.*, 2002). The process element that is, undertaking the whole project under a common/shared set of processes within a framework (the new product development process) has been looked at in the past through institutional methods of working where very often functional specialisms are promoted to the detriment of the project needs. It becomes clear that a new understanding of the new product development or design and construction process is needed to facilitate all the elements essential in delivering a project. Furthermore, sense making frameworks need to be developed that facilitate such a process. The key characteristics of a design and construction process at the project level can then be identified (Kagioglou *et al.*, 2000). Broadly speaking the design and construction process can be structured around a number of areas.

6.3.1   Pre-project stage

The pre-project stage relates to the strategic business considerations of any potential project which aims to address a client's need. There can be a number of phases within the stage and throughout those phases the client's need is progressively defined and assessed with the aim of:

- Determining the need for a construction project solution, and
- Securing outline financial authority to proceed to the pre-construction phases.

In currently acknowledged models of the design and construction process this stage of a project is given scant consideration, when compared to the latter stages of a project. However, the models assume that when approaching the Construction Industry, clients have already established 'the need'. Whilst there is little evidence to suggest this is not the case, it would seem reasonable to assume that the knowledge possessed by speculative building developers and consultants could assist any client in these early stages of a project. The problems associated with the translation of this need through the conventional briefing stage of design (O'Reilly, 1987) have the potential for substantial elimination via such an approach.

6.3.2   Pre-construction stage

With a commitment to develop the project, the process progresses through to the pre-construction stage where the defined client's need is developed into an appropriate design solution. Like many conventional models of the design process, the pre-construction phases develop the design through a logical sequence, with the aim of delivering approved production information. Given the dynamic market conditions which influence many construction client's decisions, the need for flexibility must be addressed by the industry. At the end of the pre-construction stage, all the necessary elements of a project that will enable its enactment should be in place.

6.3.3   Construction stage

The construction stage is solely concerned with the production of the project solution. It is here that the full benefits of the co-ordination and communication earlier in the process may be fully realised. Potentially, any changes in the client's requirements will be minimal, as the increased cost of change as the design progresses should be fully understood by the time on-site construction work begins.

As with all activities in the process, where concurrency is possible, it should be accommodated. This is true in particular when the complexity of the project is such that prevents or dis-carriages long term fixing of the product elements. Planning for concurrency is crucial at this stage and it should not be undertaken as an ad hoc process, either it will result in significant cost-overruns and delays.

6.3.4   Post-completion/construction stage

This is the stage where traditional projects terminate and the 'snagging' phase is initiated. A more productive approach to this stage is needed that considers the extent to which the facility is actually enabling the delivery of the business that was designed and constructed for. Furthermore, this is the stage where litigation action commences.

6.4   Key principles for an improved design and construction process

6.4.1   Sense-making frameworks that consider the whole project

In the construction industry the definition of a project has traditionally been synonymous to actual construction works. As such the pre-construction and post-construction activities have been sidelined and often accelerated to reach the construction stage or to move on to the 'new job'. This has resulted in poor client requirements identification and delayed the exposure of any potential solutions to the need to any internal and external specialists. Any contemporary attempt to define or create a 'design and construction process' will have to cover the whole 'life' of a project from recognition of a need to the operation of the finished facility and finally, to its demolition. This approach ensures that all issues are considered from both a business and a technical point of view. Furthermore this approach recognises and empha-sises the inter-dependency of activities throughout the duration of a project. It should also focus at the 'front-end' activities whereby attention is paid to the identification, definition and evaluation of client requirements in order to identify suitable solutions. Those aspects can then be used to develop frameworks and roadmaps that cover four main elements, namely:

- Structure of the work that needs to be undertaken, based on distinct project phases
- Illustration of the timing and role of the involvement of participants

- Identification of decision-making intervals aimed to ensure the quality of the product and the effectiveness of the process
- An illustration of the content of the project work within the context of the process that needs to be in place to deliver the product.

The earlier can then come together to produce sense-making frameworks and roadmaps that will engage project participants in developing a shared project understanding in terms of the work that needs to be carried out.

6.4.2   Consistency in application

A review of existing models and descriptions of the design and construction process, it can be quickly established that little consistency existed. In such an environment, the problems encountered by temporary multi-organisations (TMO) working can be compounded. Luck and Newcombe (1996) support this view, describing the 'role ambiguity' commonly associated with construction projects.

Through consistency of use of a shared process the scope for ambiguity could be reduced. This, together with the adoption of a standard approach to performance measurement, evaluation and control, should facilitate a process of continual improvement in design and construction.

6.4.3   Progressive fixity

There are problems associated with managing the unknowns of a project both in technical and business terms. An approach that promotes fixing of design and construction process elements as early as possible should increase the degree of certainty in a project. It is true however, that such certainty can rarely be achieved in construction. When this is not due to inappropriate planning, the design and construction process needs to be designed to facilitate progressive fixity through the application of CE principles and by effective planning mechanisms that aim to reduce dependency of product and service elements. Crucial to this is the availability of information at the right time to the right people and at the appropriate phase of development.

6.4.4   Co-ordination

Co-ordination is one area in which construction traditionally is perceived to perform poorly. This perception is supported by Banwell (1964), Latham (1994) and Egan (1998), in addition to many other reviews of the industry.

This is due to the fact that the co-ordination element is often not considered as an integral part of the process but rather a necessary evil. Modern approaches to partnering and framework agreements increase the degree of co-ordination simply by repeat practise, and contemporary understanding of project and production management by Koskela (1999) can improve the performance of construction projects.

6.4.5   Stakeholder involvement and teamwork

It has been recognised in many project based industries that multi-function teams, established in a development process, reduces the likelihood of costly changes and production difficulties later on in the process by enabling design and manufacturing decisions earlier in the process.

Conventionally, many building projects comprise a team of participants assembled specifically to facilitate the development of that single project. Consequently, a complete project team rarely works together on more than one project, and, as Sommerville and Stocks (1996) argue, this can negatively affect the assembled 'team's' performance. In addition, many key contributors are identified and included too late in the process.

Project success relies upon the right people having the right information at the right time. Proactive resourcing of project phases through the adoption of a 'stakeholder' view should ensure that appropriate participants (from each of the key functions) are consulted earlier in the process than is traditionally the case. This, in itself, will not eliminate the problems associated with TMO working. However, the active involvement of all participants, especially in the early phases of a project, may subsequently help foster a team environment and encourage appropriate and timely communication and decision-making.

6.4.6   Customisation and flexibility

A process is by definition evolving through time and according to lessons learned. As such, any design and construction process should allow an element of flexibility in terms of the content of the work and the way in which project participants interact within this process. For example, the procurement routes chosen by the project consortium will largely determine the point at which participants enter the process and their role within the project. All procurement routes have their pros and cons but provisions should be made to accommodate their enactment. It can be argued that different design and construction processes should exist for different types of procurement, and possibly the same can be argued for different building types and so on. As the number of project specific variables increases however so does the degree of complexity and the number of different types of processes. Rather a more generic approach should be adopted that facilitates flexibility and customisation. Therefore, the design and construction process should be independent of procurement routes and other variables, but rather designed to accommodate such project specific issues.

6.4.7 Feedback

In addition to the direct teamwork problems associated with TMO's, the ability to learn from experience is also hampered by the continual formation and break-up of project teams. Both success and failure can offer important lessons for the future, yet the fragmented and competitive nature of the construction industry prevents the benefits of shared good practice being utilised. The design and construction process should facilitate a means by which project experiences can be recorded, throughout the process, thereby informing later phases and future projects. Competitive advantage will come from how such experiences are acted upon. Shared knowledge may not automatically increase the competitiveness of companies working in construction, however, the subsequent increase in awareness, project to project, has the potential for reducing risk and improving performance.

6.5 The design and construction Process Protocol

The Generic Design and Construction Process Protocol (GDCPP) was developed by the University of Salford in an attempt to improve the delivery process of construction projects. It is presented in a high-level process map that aims to provide a framework to help companies achieve an improved design and construction process. The map draws from principles developed within process management field that include stakeholder involvement, teamwork and feedback, and reconstructs the design and construction team in terms of Activity Zones rather than in disciplines to create a cross-functional team. These Activity Zones are multi-functional and may consist of a network of disciplines to enact specific task of the project, allowing the 'product' to drive the process rather than the function as in a sequential approach. The use of zones potentially reduces this confusion and enhances communication and co-ordination (Kagioglou et al., 1998). The Activity Zones contain high-level processes spanning the duration of a project from inception, through design and construction, and including operation and maintenance. The responsibility for completing the processes may lie with one Activity Zone or be shared.

Furthermore, the Process Protocol aims 'to map the entire project process [NPD] from the client's recognition of a need to operations and maintenance' (Kagioglou et al., 2000). The protocol takes the form of a framework detailing the generic design and construction processes within a construction project. The intention was for construction firms to take the map and to use it as a framework to help them to improve their business and through industry interest and acceptance, further funding has been committed to continue the research. It was envisaged that the generic protocol would not be an ad hoc activity, but an ongoing and planned one.

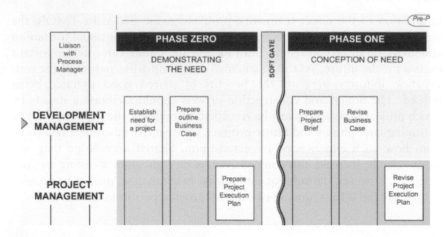

Figure 6.3 Process Protocol framework.

Therefore, the framework should not be so prescriptive as to restrict or stifle creativity but be easily adapted and tailored to suit the individual project. This brings the generic protocol down to a secondary-level (Level 2) or product-specific level, which itself can be broken down further to more detailed levels to create sub process maps of the eight Activity Zones within the Generic Design and Construction Process Protocol Model (see Figure 6.3 for a short illustration). The Process Protocol Level II project[1] subsequently aimed to identify such sub processes, however, the implementation of the framework (Kagioglou *et al.*, 2002) also highlighted some issues:

- Due to the complexity of the construction project, the process model will become very complicated. It is almost impossible to manage all the processes manually.
- Companies might only adopt part of the Process Protocol model, depending on the nature of the project.
- Some companies have their own working process and are not willing or able to accommodate a new approach.
- The individuals who are responsible for the process modelling and management of a project need detailed knowledge of the Process Protocol.
- The opportunities presented by Internet technology for organisations to improve the performance and more effectively reach the parties involved in the project is now being used and the Process Protocol needs to adapt to the technology.

An IT solution, the Process Protocol toolkit, is needed to resolve theses issues. It is being developed under the Process Protocol Level II project.

The tool aims to assist the creation of the process model and to manage the processes based on the Process Protocol framework, and will be discussed in detail later in this chapter.

6.5.1   Process Protocol framework

The Process Protocol framework consists of the following major elements (see Kagioglou *et al.*, 1998 for a detailed description).

6.5.1.1   Process

A set of activities undertaken by multifunctional team is to produce information for other processes or deliverables. For example, 'establish need for project'.

6.5.1.2   Deliverable

As output of the process, deliverables represent documented project and process information, such as Stakeholder List, Statement of need, project brief, etc.

6.5.1.3   Phase

There are 10 phases that have been defined in the Process Protocol map to represent the different stage of the whole lifecycle of a construction project.

6.5.1.4   Activity zone

Nine Activity Zones in the Process Protocol map represent the different group of participants involved in a construction project, namely Development Management, Project Management, Resource Management, Design Management, Production Management, Facilities Management, Health and Safety, Statutory and Legal Management and Process Management.

6.5.1.5   Phase reviews

They are conducted by a multifunctional senior management group and representatives of the project team. The work is in the form of deliverables as described in the Process Protocol and are assessed in the Phase Review meeting. The Phase Review report will include key deliverables for the appropriate phase as identified by the project process map.

6.5.2   Process representation

The processes and sub-processes are denoted by using the symbol shown in Figure 6.4.

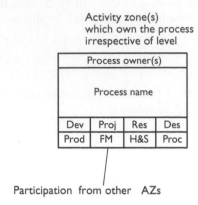

Activity zone(s)
which own the process
irrespective of level

Figure 6.4 Process representation.

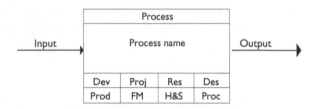

Figure 6.5 Inputs and outputs to the process.

6.5.2.1   Process owner(s)

Process name (potentially including some description for clarification where required). An indication of likely/potential participation from other activity zones in the process.

6.5.2.2   Inputs

For clarity, inputs to a process are only shown where they form a logical dependency from another process at that level on the same diagram (see Figure 6.5). All other inputs from different phases or Activity Zones are not shown, but are traceable through the modelling database.

6.5.2.3   Outputs and deliverables

All processes by definition have an output. Some of these can be called 'deliverables', where the information is in a form (or document) that should be named for easy reference and use in other processes. The outputs to be named as deliverables are defined later in the Process Protocol framework.

6.5.2.4   Process Levels

The maps contain three levels:

- Level I contains the high level processes and their deliverables as identified in the Process Protocol Map.
- Level II contains the sub-processes of the main process at Level I (i.e. what the Level I process consists of) and how those sub-processes interact with each other (i.e. how is the Level I process undertaken).
- Level III contains the sub-processes of the processes at Level II (what the Level II process consists of) and how those sub-processes interact with each other (how is the Level II process undertaken).

6.6   Process Protocol toolkit

Having described the Process Protocol framework the way in which the Process Protocol toolkit can support the Process Protocol mapping and its principles can be outlined. The Process Protocol toolkit is composed of two major components; process map creation tool and process management tool. To develop this toolkit, it is vital to understand the information relationships between the major elements of the Process Protocol framework. Data model of the Process Protocol framework was produced to illustrate the relationships.

The methodology for the data modelling is Entity Relationship Diagram (ERD), which was introduced in the 1970's by Peter Chen to model the design of a relational database from a more abstract perspective. (Chen, 1997).

An Entity relationship diagram (ER diagram) uses three major abstractions to describe the data. They are:

- Entities, which are distinct and major elements in the business; that is, map element 'activity zone'.
- Relationships, which are meaningful interactions between the entities; that is, entity 'activity zone' and entity 'process', the relationship between is 'One activity zone has one or more processes'.
- Attributes, which are the properties of the entities and relationships, that is, name, description of entity 'activity zone'.

The entity relationship diagram in Figure 6.6 represents how the major elements of the Process Protocol framework are interacted each other and how the information associated with them can be stored.

This ERD model is turned into a database by mapping the entities and relationships as database tables to hold the data of project process map created by the process map creation tool.

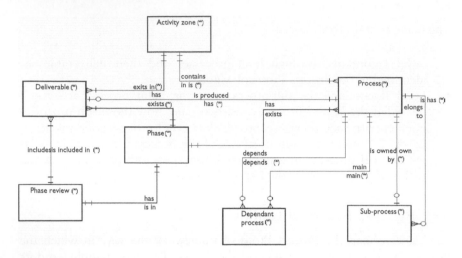

Figure 6.6 Entity relationship diagram for Process Protocol framework.

6.6.1 *Process map creation tool*

The process map creation tool is a process-mapping tool specially designed for the creation of the project process map based on the Process Protocol framework. It automates the map creation process and guides the user who might lack of the knowledge of the Process Protocol to create a project process map at early stage of the project. Users will be able to tailor and customise the process map to suit their own project and company requirements.

To some extent, the process map creation tool is very similar as process modelling tools that have been available on the software market for years. Many companies have adopted a process-oriented view of their business operation, replacing the traditional functional viewpoint to achieve a better integration of operation (Hammer and Champy, 1995). Therefore, software tools to assist such approach have been developed and they can be categorised into two major types, paper based diagramming tools and software enabled analysis tools.

Paper based diagramming tools primarily offer the integration of diagrams and illustrations, together with a wide variety of other features and abilities. Most of the tools provide drawing support with templates or shapes, which can be customised to suit individual requirements. The industry standard modelling languages, such as Integrated Computer Aided Manufacturing Definition (IDEF), Data Flow Diagram, Entity Relationship Diagram, have been incorporated into these products.

Software enabled analysis tools are more commonly called BPR tools or Computer Aided Software Engineering (CASE) tools and usually encompassed built-in event simulator, static analysis, dynamic modelling and standard database support. These tools are able to produce a descriptive model that attempts to represent the business 'as is' or 'as to be'. Such a model can be composed of a number of process definitions including goals, business rules, actions and resource requirements, and expresses the flow of activity between the processes with a combination of diagrams, text and performance measures. Typically, the business model is built using a process modeling (built-in) tool, and they then may simulate the running of the process. However, most tools focus on IDEF methodology and several are based on the Data Flow or Entity Relationship Diagram. Although these process tools provide powerful functions, they cannot be effectively used as an IT support for the Process Protocol, because the aim of the toolkit is to help the industry implement the Process Protocol and not to analyse the construction process. In addition, the Process Protocol has its own process modeling methodology which was developed with the industry to facilitate their own simple requirements, though this is not discussed in the extent of this chapter. All of the intelligent tools only support standard accepted modeling methodologies, like IDEF, data flow diagram and therefore, the Process Protocol toolkit needs to be developed to fulfill the role in the project.

The prototype of the process map creation tool has been developed under the Process Protocol II project. It enables the production of a project process map based on the generic Process Protocol framework. There are three major components in the tool, which are main creation tool, generic processes data store and project process data store.

The main creation tool provides the functions for data retrieval, map creation and map customisation. Users will be able to define their processes, and create the project process map by referring to the generic processes provided by Process Protocol. All the generic processes developed in the Process Protocol project are stored in the generic process data store that has been built according to the Process Protocol data model. The project process map created by users is stored in the project process data store, which becomes the basis of the process management tool.

Figure 6.7 is a screen shot of the prototype of the process map creation tool. It is a standalone MS Windows application developed using Microsoft Visual Basic programming tool. Its interface consist of three main parts.

6.6.1.1   Process tree

On the left side of the window, Process tree is used a similar windows file explorer style to show the decomposition structure of the process map. Processes in three different levels represent in process tree hierarchy respectively. Processes in the process tree are selectable, they can be selected by mouse click and the corresponding process in process map will be highlighted. In Figure 6.7, the process 'identify space requirements' is selected and the same process in process map is highlighted.

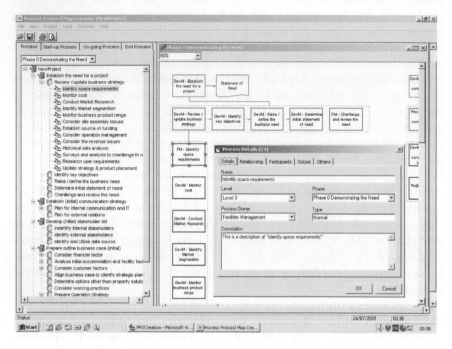

Figure 6.7 Process map creation tool.

6.6.1.2   Process map

Process map is a visual representation of the Process Protocol map, it interacts with the process tree on the left. Processes in different levels are represented in different colours. In this case, Process 'identify space requirements' is level 3 process and it is in blue.

6.6.1.3   Process details

All the information associated with each process is shown in process details dialogue box. It includes name, process level, process owner, description, type, etc.

6.6.2   Process management tool

The process management tool is a web based project information management system by integrating the process as a core information structure. It provides functionalities for project management and workgroup collaboration in a virtual environment, such document sharing, document and

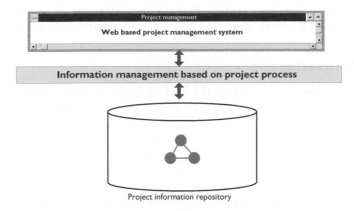

Figure 6.8 Process management tool system architecture.

drawing management, online publishing, user control, etc. Project teams can secure and centralise the engineering and project information for all that need to see it. In this environment, teams can reduce costs and save time as they gather and disseminate information throughout the project lifecycle. Furthermore, the integrated project process map will become the route map to help and guide the project management team to monitor and track project progress, documents, etc. The centralised the project information can be reused in the future project as reference.

The proposed process management tool stores all the project document and information according to the project process, which is created by the process map creation tool. The project process effectively becomes the information structure of the project. Users are still able to search the information in traditional way, but more important, users can follow the project process to locate the information they might need. That is major difference between the process driven management tool and current project extranet.

The proposed system architecture of process management tool is presented in Figure 6.8. It is composed of three layers, they are the web based project collaboration system layer, the information Management based on project process layer and the project information repository layer. The web based project management system provides all the usual functionalities, such as document management, user control, messaging service and collaboration service. It also has an interface for viewing project process maps, navigating project process. It is front-end the process management tool to guide the users to manage the project process and project information. The information management layer includes project process information created in process map creation tool. It is a mechanism to archive the project information according to the project process. It provides data management facilities for the project information repository. The project information repository is a database system to hold the information of the project, including the document, drawings, program information, cost data, etc.

6.7   Conclusions

This chapter aimed to introduce the field of design and construction process management for concurrent lifecycle. It has identified and presented the key principles that should be encompassed within an improved process and demonstrate those through the description of the Process Protocol.

The field of process management within the construction sector has enjoyed some progress in the last few years but there is still considerable research that needs to be carried out. For example, considering construction as a complex, chaotic system as suggested by Bertelsen (2004) and try to manage it as such can prove to have significant importance for future practices. Also, contemporary understanding of project and production management promoted by the International Group for Lean Construction (IGLC) can prove to be the catalyst in new paradigms been developed that provide a more realistic picture of the sector.

Furthermore, considerable research effort is needed to enable the implementation of new paradigms and to develop a base of empirical evidence to support those paradigms.

6.8   Note

1 In collaboration with Loughborough University and eleven industrial partners.

6.9   References

Anumba, C. J., Baugh, C. and Khalfan, M. M. A. (2002). 'Organisational Structures to Support Concurrent Engineering in Construction', *Industrial Management & Data Systems*, Vol. 102, Part 5, pp. 260–270.

Banwell, H. (1964). 'Report of the Committee on the Placing and Management of Contracts for Building and Civil Engineering Works', HMSO.

Bertelsen, S. (2004). 'Construction Management in a Complexity Perspective', The 1st SCRI International Symposium, University of Salford, UK.

Bulletpoint (1996). 'Creating a Change Culture – not about Structures, but Winning Hearts and Minds'. Sample issue, pp. 12–13 (in Zairi, M. (1997)).

Chen, P. (1997). 'The Entity – Relationship Model – Toward a Unified View of Data', *ACM Transactions on Database Systems* 1.

Davenport, T. H. (1993). *Process Innovation: Reengineering Work Through Information Technology*, Harvard Business School Press, Boston, MA, USA.

Egan, J. (1998). 'Rethinking Construction', Department of the Environment, Transport and the Regions.

Elzinga, D. J., Horak, T., Chung-Yee, L. and Bruner, C. (1995). 'Business Process Management: Survey and Methodology', *IEEE Transactions on Engineering Management*, Vol. 24, No. 2, pp. 119–128.

Hammer, M. and Champy, J. (1995). *Reengineering the Corporation: A Manifesto for Business Revolution*, Nicholas Brealey Publishing Limited, London, UK.

Harrington, H. J. (1991). *Business Process Improvement*, McGraw Hill, New York, NY.

Kagioglou, M., Cooper, R., Aouad, G., Hinks, J., Sexton, M. and Sheath, D. (1998). 'A Generic Guide to the Design and Construction Process Protocol', University of Salford, UK.

Kagioglou, M., Cooper, R., Aouad, G. and Sexton, M. (2000). 'Rethinking Construction: The Generic Design and Construction Process Protocol', *Journal of Engineering Construction and Architectural Management*, Vol. 7, No. 2, pp. 141–154.

Kagioglou, M., Lee, A., Cooper, R., Carmichael, S. and Aouad, G. (2002). 'Mapping the Construction Process: A Case Study', 10th International Conference of Lean Construction, Gramado, Brazil, 6–8 August.

Koskela, L. (1999). Management of Production in Construction: A Theoretical View. In Proceedings of the 7th International Group of Lean Construction, University of California, Berkeley, USA, 241–252.

Latham, M. (1994). '*Constructing the Team*', HMSO.

Lee, R. G. and Dale, B. G. (1998). 'Business Process Management: A Review and Evaluation', *Business Process Management Journal*, Vol. 4, No. 3, pp. 214–225.

Luck, R. and Newcombe, R. (1996). 'The Case for the Integration of the Project Participants' Activities within a Construction Project Environment', in Langford, D. A. and Retik, A. (eds), *The Organization and Management of Construction: Shaping Theory and Practice (Volume Two)*; E.&F.N. Spon.

Oakland, J. S. (1995). *Total Quality Management: The Route to Improving Performance*, Butterworth Heinemann Ltd., Second Edition.

O'Reilly, J. J. N. (1987). '*Better Briefing means Better Buildings*', Building Research Establishment report BR 95 B.R.E., Garston, UK.

Ould, M. A. (1995). *Business Processes: Modelling and Analysis for Reengineering and Improvement*, John Wiley & Sons, Chichester.

Peppard, J. and Rowland, P. (1995). *The Essence of Business Process Re-Engineering*, Prentice Hall.

Rosenau, M. (1996). 'The PDMA Handbook of New Product Development', John Willey & Sons, Inc., USA.

Sommerville, J. and Stocks, B. (1996). 'Realising the Client's Strategic Requirements: Motivating Teams', Proceedings of COBRA'96, University of the West of England.

Talwar, R. (1993). 'Business Re-engineering – a Strategy-Driven Approach', *Long Range Planning*, Vol. 26, No. 6, pp. 22–40.

Zairi, M. (1997). 'Business Process Management: a Boundaryless Approach to Modern Competitiveness', *Business Process Management Journal*, Vol. 3, No. 1, pp. 64–80.

Chapter 7

Ontologies and standards-based approaches to interoperability for Concurrent Engineering

Line C. Pouchard and Anne-Francoise Cutting-Decelle

7.1 Introduction

As the use of information technology and computer-driven systems in manufacturing and construction design has matured, the necessity for software applications to work together, exchange data, processes and information has become crucial to the conduct of business and operations in organisations. This capability is referred to as interoperability (Pouchard, 2000, 2002) (Ray, 2003). To be competitive and maintain good economic performance, organisations need to employ increasingly effective and efficient data and computer systems. Such systems should result in the seamless integration of application data and exchange of processes between applications. Organisations should also be able to conserve and retrieve on demand the knowledge contained in their business and operational processes, regardless of the applications used to produce and handle these processes.

With the increasing need for enterprise integration, developers face more complex problems related to inter-operability. Independent contractors and suppliers who collaborate on demand within virtual supply chains must share product-related data. Vendor applications that are not designed to inter-operate must now share processes. When enterprises collaborate, a common frame of reference or at least a common terminology is necessary for human-to-human, human-to-machine and machine-to-machine communication. Similarly, within a core enterprise where distributed collaboration between remote sites and production units take place, a common understanding of business- and manufacturing-related terms is indispensable. However, this common understanding of terms is often at best implicit in the business transactions and software applications and may not even be always present. Misunderstanding between humans conducting business-related tasks in teams, and ad-hoc translations of software applications contribute to the rising costs of interoperability in manufacturing.

Standard-based approaches and ontologies offer a direction addressing the challenges of interoperability brought about by semantic obstacles, that is, the obstacles related to the definitions of business terms and software classes. An ontology is a taxonomy of concepts and their definitions supported by a logical theory (such as first-order predicate calculus). Ontologies have been defined as an explicit specification of a conceptualisation (Gruber, 1993). An ontology expresses, for a particular domain, the set of terms, entities, objects, classes and the relationships between them, and provides formal definitions and axioms that constrain the interpretation of these terms (Gomez-Perez, 1998). An ontology permits a rich variety of structural and non-structural relationships, such as generalisation, inheritance, aggregation and instantiation and can supply a precise domain model for software applications (Huhns and Singh, 1997). For instance, an ontology can provide the object schema of object-oriented systems and class definitions for conventional software (Fikes and Farquhar, 1999). Ontological definitions, written in a human readable form, can be translated into a variety of logical languages. They can also serve to automatically infer translation engines for software applications. By making explicit the implicit definitions and relations of classes, objects and entities, ontology engineering contributes to knowledge sharing and re-use (Gomez-Perez, 1998). Ontology engineering aims at making explicit the knowledge contained within software applications, and within enterprises and business procedures for a particular domain and includes a set of tasks related to developing ontologies for a particular domain.

Interoperability in manufacturing refers to the ability to share technical and business information seamlessly throughout an extended enterprise (supply chain) (Ray and Jones, 2003). This information, previously shared in a variety of ways including paper and telephone conversations, must now be passed electronically and error-free with suppliers and customers around the world. A study, achieved by the NIST in 2002 (NIST, 2002) was aimed at identifying the economic impact of the use of standards in industry, particularly the ISO 10303 STEP standard with the objective of conducting an economic impact assessment of STEP's use by transportation equipment industries, namely the automotive, aerospace, shipbuilding and specialty tool and die industries. Both the full potential and current realised benefits are quantified. In addition, the study investigates the impact of NIST's administrative and technical contributions to STEP. The authors of the study estimate the economic value of the efficiency gains due to improved data exchange enabled by using STEP, and quantify NIST's contributions to those gains. Data collected from industry surveys and case studies are used to estimate the potential benefits of existing STEP capabilities. They estimate that STEP has the potential of save $928 million (in 2010) per year by reducing interoperability problems in the automotive, aerospace, and shipbuilding industries. Currently approximately 17 per cent ($156 million) of the potential benefits of STEP quantified within the scope of this study are being realised. A previous study commissioned by NIST (NIST, 1999) in 1999, had reported that the US automotive sector alone expended one billion dollars per year to resolve interoperability problems. The study also reported that as much as 50 per cent of this expenditure is attributed to dealing with data file exchange issues.

7.2 Interoperability in construction, what do we mean?

7.2.1 Information systems in construction, specificity and main features

Information systems become increasingly important in industrial companies for acquiring, structuring and exchanging complex technical data that they have to handle during the production process. The intrinsic complexity of the information becomes yet more complex with the relational structuring of the data. This structure is necessary in order to select among the set of possible solutions the most competitive ones in answer to given specifications. This is particularly true for construction SMEs, since they are often exposed to situations for which they have neither the necessary skills nor the tools enabling a continuous updating of the technical information needed by the projects they work on and the software tools they use (Cutting-Decelle, Dubois, in Bestougeff *et al.*, 2002).

Fundamentally, the construction industry is characterised by:

- an increasing complexity with an acceleration of the relations among the partners, particularly in a CE context, alongside a dramatic reduction of the lead-time between the call for tender and the operation of the building (Anumba *et al.*, 1999);
- an increasing diversity of the information and data handled, mainly due to the development of new representation structures (use of standard messages such as EDIFACT messages, use of product data- de facto or de jure-standards: STEP, P-LIB and IFCs, as we will see in this chapter;
- the development of new software tools capable of dealing with the increasing volume and diversity of information, although, most of the time, without any interoperability between them;
- a great heterogeneity of the information handled, since a normal construction project requires several documents simultaneously. These include drawings, calculation, technical notes, bills of materials and other kinds of technical analysis, as well as documents (legal or not) containing information related to the different building components.

The evaluation of the degree of elaboration of an information system starts with the possibility to identify and to interface, when possible, existing document repositories or product databases, regardless of their structuring and location.

7.2.2   The concept of interoperability

Interoperability is 'the ability of software and hardware on multiple machines from multiple vendors to communicate' (FOLDOC). It is also considered as 'the ability of a system or a product to work with other systems or products without special effort on the part of the customer. Interoperability becomes a quality of increasing importance for information technology products as the concept that "The network is the computer" becomes a reality' (whatis.com).

According to Miller (2000), 'to be interoperable, one should actively be engaged in the ongoing process of ensuring that the systems, procedures and culture of an organisation are managed in such a way as to maximise opportunities for exchange and re-use of information, whether internally or externally'.

Based upon this definition, it should be clear that there is far more to ensuring interoperability than using compatible software and hardware. Rather, assurance of effective interoperability will require often radical changes to the ways in which organisations work and, especially, in their attitudes to information.

Different approaches to the challenges of interoperability exist. One is likely to find them in combination in real-world problems:

- A standard-based approach: the most straightforward aspect of maintaining interoperability. Consideration of technical issues includes ensuring an involvement in the continued development of communication, transport, storage and representation standards. Work is required both to ensure that individual standards move forward to the benefit of the community, and to facilitate where possible their convergence, such that systems may effectively make use of more than one standards-based approach.
- A software engineering approach: In this approach, software developers and quite often users who need the data outputs of an application as input to another write some syntactic parsers that allow the language and/or the data structures of the output to be mapped to the structures and language of the second application. This approach does not take into account the semantic conflicts and gaps described later. Furthermore, the mappings between two applications are ad hoc, that is left to the subjective understanding of concepts by developers. Finally, each time a new application or even a new version of an existing application occurs, the parsers need to be modified.
- A semantic interoperability approach: Semantic interoperability presents a host of issues, all of which become more pronounced as

individual resources – each internally constructed in their own
semantically consistent fashion – are made available through 'gate-
ways' and 'portals'. Almost inevitably, these discrete resources use different terms to describe similar concepts ('slab', 'floor', 'level', 'surface',
for example), or even use identical terms to mean very different things,
introducing ambiguïty and error into their use. This situation is trou-
bling because the errors introduced are not necessarily explicit and may
induce errors in analysis or design.

There are also other kinds of interoperability, among which we will
mention: human interoperability, inter-community interoperability and legal
interoperability. In this chapter, we will focus on technical interoperability
among software tools used by the professionals of the construction sector.

7.2.3   The need for interoperability

Being seen to 'be interoperable' is becoming increasingly important to a
wide range of organisations, projects, even companies. In each case unde-
niably valuable information is being made available to a wide range of
users, often for the first time. The drive towards interoperability will necessarily lead to changes in the way the organisations operate. One of
the aims of this book is to show that concurrent engineering provides a
valuable tool of the interoperability.

A truly interoperable organisation is able to maximise the value and reuse
the potential of information under its control. It is also able to exchange
this information effectively with other equally interoperable bodies, allowing new knowledge to be generated from the identification of relationships
between previously unrelated sets of data.

The lack of interoperability is very costly to some industrial sectors. But
changing internal systems and practices to make them interoperable is a
non-trivial task. However the benefits for the organisation and those
making use of information it publishes are potentially incalculable, as
mentioned in the introduction to this chapter.

7.2.4   The potential of standards to increase
interoperability

There are three principal approaches to compensate for the lack of
interoperability:

The first is a point-to-point customised solution, which can be achieved
by contracting the servïces of systems integrators. This approach is expensive
since each pair of systems needs a dedicated solution.

A second approach, adopted in some large supply chains, requires all
partners to conform to a particular solution. This approach does not solve

the interoperability problem since the first or sub-tier suppliers are forced to purchase and maintain multiple, redundant systems. It can also be costly to the smaller organisations in a supply chain since they are rarely in a position to influence the choice of infrastructure, and may not have enough resources to comply.

The third approach involves neutral, open, published standards. By adopting open standards the combinatorial problem is reduced from n^2 to n, with bi-directional translators.

Published standards also offer some stability in the representation they propose of the information models, an essential property for long-term data archiving. This chapter highlights some of the standards developed within the ISO TC184 'Industrial Automation Systems and Integration' Committee, particularly those relevant to the construction sector (ISO).

But the problem is far from solved. Interoperability standards are used in layers, from the cables and connectors, through networking standards, to the application or content standards such as those mentioned here, that is STEP, P-LIB and PSL (Process Specification Language). All of these layers must function correctly for interoperability to be achieved. The greatest challenges remain at the top of this stack of standards, in order to make them interoperable. Due to the capability of the PSL language to be extended (through its ontology) for accommodating concepts in other standards, this language can be considered as a powerful tool of this interoperability, enabling, for a near future, the consideration of a 'universal interfacing'.

We present in the following sections some of the main (de facto and de jure) standards that can be used in construction. Since this approach of the construction sector with an interoperability based on standards is rather new, we describe the most known in the domain of product data modelling (ISO 10303 STEP, ISO 13584 P-LIB and IAI/IFCs) but also a new standard used for the specification of process related information, the ISO 18629 PSL standard. This PSL language brings an important contribution to the problem of the semantic ambiguity met in the information exchanges.

7.3   International standards developed by the ISO TC184 committee

The ISO TC184 is one of the one two hundred committees managed by the International Standardisation Organisation, Geneva, CH (ISO), its scope is: 'Standardisation in the field of industrial automation and integration concerning discrete part manufacturing and encompassing the applications of multiple technologies, that is, information systems, machines and equipments and telecommunications.' This means that the standards developed are applicable to manufacturing and process industries, applicable to all sizes of business, applicable to extending exchanges across the globe through e-business.

Excluded from the scope are the following domains: electrical and electronic equipment (dealt with by the IEC/TC44) and programmable logical controllers for general applications (IEC/TC65). The scope of the committee means that the standards developed are: applicable to manufacturing and process industries, applicable to all sizes of business, applicable to extending exchanges across the globe through e-business.

The standards developed within the ISO TC184 cover various domains related to industrial automation and integration, among which: enterprise modelling, enterprise architecture, communications and processes, integration of industrial data for exchange, access and sharing, life cycle data for process plants, manufacturing management, mechanical interfaces and programming methods, part libraries, physical device control, Process Specification Language (PSL), product data and robots for manufacturing environment (Cutting-Decelle et al., 2004).

7.3.1   ISO 10303 STEP

Each part of ISO 10303 contains the following introductory paragraph that summarises the significant challenges undertaken in this standardisation effort (Kemmerer, 1999):

> ISO 10303 is an International Standard for the computer-interpretable representation and exchange of product data. The objective is to provide a neutral mechanism capable of describing product data throughout the lifecycle of a product, independent from any particular system. The nature of this description makes STEP suitable not only for neutral file exchange, but also as a basis for implementing, sharing product databases, and archiving.
>
> (IS 10303–1, 1994)

STEP was designed to be the successor of exchange standards such as IGES, SET and VDA-FS with the notable difference that it was intended to do more than support exchange of product data. STEP is intended to support data sharing and data archiving. These distinguishing concepts are given below:

Product data exchange   The transfer of product data between a pair of applications. STEP defines the form of the product data that is to be transferred between a pair of applications. Each application holds its own copy of the product data in its own preferred form. The data conforming to STEP is transitory and defined only for the purposes of exchange.

Product data sharing   The access of and operation on a single copy of the same product data by more than one application, potentially simultaneously. STEP is designed to support the interfaces between the single copy of the product data and the applications that share it. The applications do not hold the data in their own preferred forms. The architectural elements of STEP may be used to support the realisation of the shared product data itself. The product data of prime interest in this case is the integrated product data and not the portions that are used by the particular product data applications.

Product data archiving   The storage of product data, usually long term. STEP is suitable to support the interface to the archive. As in product data sharing, the architectural elements of STEP may be used to support the development of the archived product data itself. Archiving requires that the data conforming to STEP for exchange purposes is kept for use at some other time. This subsequent use may be through either product data exchange or product data sharing.

Early in the development of ISO 10303, SC4 recognised that the scope of the standard was extremely large. This fact resulted in a couple of fundamental assumptions that shaped the architecture of STEP. SC4 assumed it unlikely that any one organisation would implement the entire ISO 10303, due to its large scope. Therefore, it made sense to separate the standard into parts, where an organisation would implement only the subset of parts needed to satisfy the requirements of their operation. Another primary concept contributing to the architecture is that the content of the standard is to be completely driven by industrial requirements. This, in combination with the concept that the re-use of data specifications is the basis for standards, led to developing two distinct types of data specifications. The first type, reusable, context independent specifications, defines the building blocks of the standard. The second type, application-context-dependent specifications (application protocols) is developed to satisfy clearly defined industrial information requirements. This combination enables avoiding unnecessary duplication of data specifications between application protocols.

SC4 determined that computer-sensible standards specifications were necessary to facilitate reliability and efficiency. The expression of STEP data constructs through a formal data definition language is necessary (but not sufficient) for unambiguous definition of data.

7.3.1.1   Components of ISO 10303

The architecture of STEP is intended to support the development of standards for product data exchange and product data sharing. The requirements and concepts in the preceding section have contributed to the evolution of the architecture over the past decade. The architectural components of STEP are reflected in the decomposition of the standard into several series of parts. The STEP document composition was developed at the June 1989 meeting of ISO TC184/SC4/WG1 as a series of parts. Each part series contains one or more types of ISO 10303 parts. Figure 7.1 provides an overview of the structure of the STEP documentation.

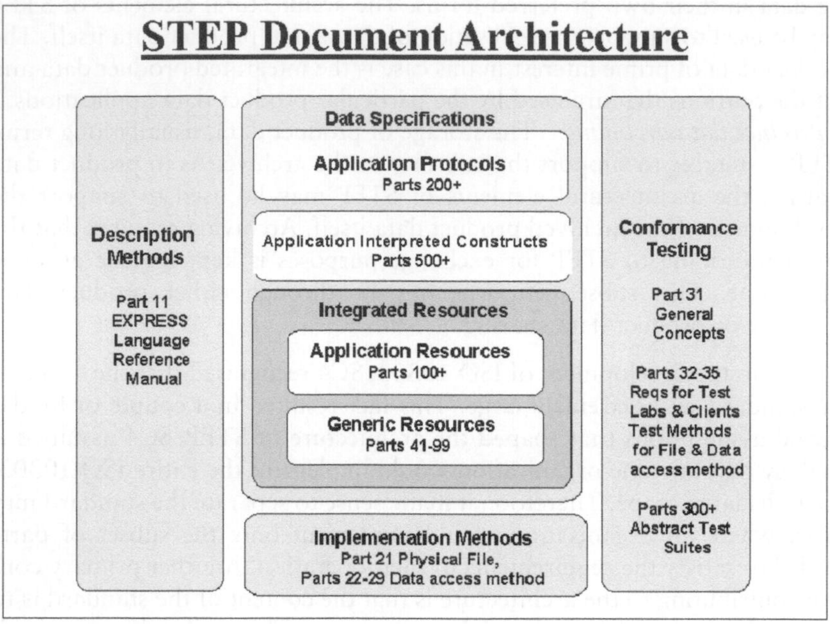

Figure 7.1 Overview of the STEP document architecture.

Source: Kemmerer (1999).

The following describes each of the structural components and functional aspects as an overview of the STEP architecture.

Description methods   The first major architectural component is the description method series of STEP parts. Description methods are common mechanisms for specifying the data constructs of STEP. Description methods include the formal data specification language developed for STEP, known as EXPRESS (10303–11, 2004). Other description methods include a graphical form of EXPRESS, a form for instantiating EXPRESS models, and a mapping language for EXPRESS.

Implementation methods   The second major architectural component of STEP is the implementation method series of 10303 parts. Implementation methods are standard implementation techniques for the information structures specified by the only STEP data specifications intended for implementation, application protocols. Each STEP implementation method defines the way in which the data constructs specified using STEP description methods are mapped to that implementation method. This series includes the physical file exchange structure (10303–21, 1994), the standard data access interface (10303–22), and its language bindings (10303–23, –24, –26). Implementation methods are standardised in the ISO 10303–20 series of parts.

Conformance testing The third major architectural component of STEP is in support of conformance testing. Conformance testing is covered by two series of 10303 parts: conformance testing methodology and framework, and abstract test suites. The conformance testing methodology and framework series of 10303 parts provide an explicit framework for conformance and other types of testing as an integral part of the standard. This methodology describes how testing of implementations of various STEP parts is accomplished.

Data specifications The final major component of the STEP architecture is the data specifications (see Figure 7.2). There are four part series of data specifications in the STEP documentation structure, though conceptually there are three primary types of data specifications: integrated resources, application protocols and application interpreted constructs. All of the data specifications are documented using the description methods.

Integrated resources The integrated resources constitute a single, conceptual model for product data. The constructs within the integrated resources are the basic semantic elements used for the description of any product at any stage of the product lifecycle. Although the integrated resources are used as the basis for developing application protocols, they are not intended for direct implementation. They define reusable components intended to be combined and refined to meet a specific need. The integrated resources comprise two series of parts, the integrated generic resources and the integrated application resources. The two series have similar function and form: they are the application, context-independent standard data specifications that support the consistent development of application protocols across many application contexts.

Application protocols Application protocols (APs) are the implementable data specifications of STEP. APs include an EXPRESS information model that satisfies the specific product data needs of a given application context. APs may be implemented using one or more of the implementation methods. They are the central component of the STEP architecture, and the STEP architecture is designed primarily to support and facilitate developing APs. Many of the components of an application protocol are intended to document the application domain in application specific terminology. Application protocols are standardised in the ISO 10303–200 series of parts.

Application interpreted constructs Application interpreted constructs (AICs) are data specifications that satisfy a specific product data need that arises in more than one application context. An application interpreted construct specifies the data structures and semantics that are used to exchange product data common to two or more application protocols. Application protocols with similar information requirements are compared semantically

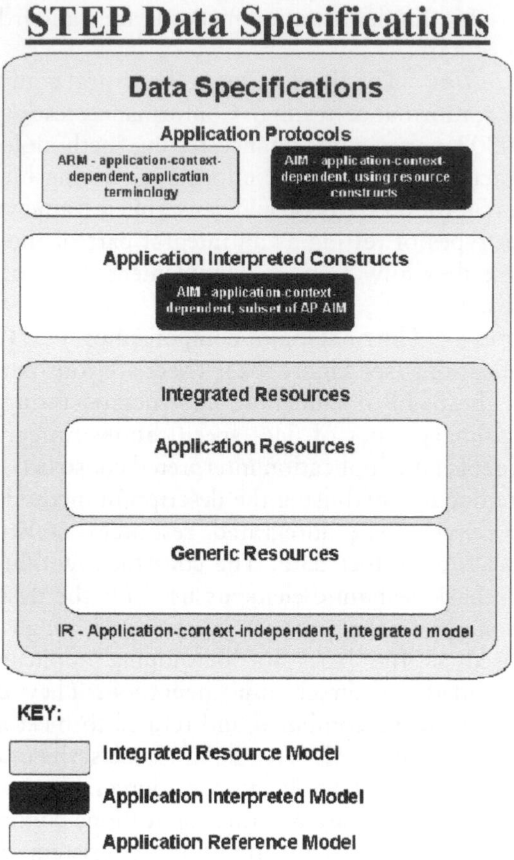

STEP Data Specifications

Data Specifications

Application Protocols

| ARM - application-context-dependent, application terminology | AIM - application-context-dependent, using resource constructs |

Application Interpreted Constructs

AIM - application-context-dependent, subset of AP AIM

Integrated Resources

Application Resources

Generic Resources

IR - Application-context-independent, integrated model

KEY:

Integrated Resource Model

Application Interpreted Model

Application Reference Model

Figure 7.2 STEP data specification.

Source: Kemmerer (1999).

to determine functional equivalence that, if present, leads to specifying that functional equivalence within a standardised AIC. This AIC would then be used by both application protocols and available for future APs to use as well. STEP has a requirement for interoperability between processors that share common information requirements. A necessary condition for satisfying this requirement is a common data specification. Application interpreted constructs provide this capability. Application interpreted constructs are standardised in the ISO 10303–500 series of parts.

For a few years, a new concept, called 'common resources' has appeared within the STEP community. This concept is aimed at maximising the re-use of already existing elements, either directly within the data specifications, or by means of the development of 'application modules': see Figures 7.3, 7.4 and 7.5 for further information about the components of the standard.

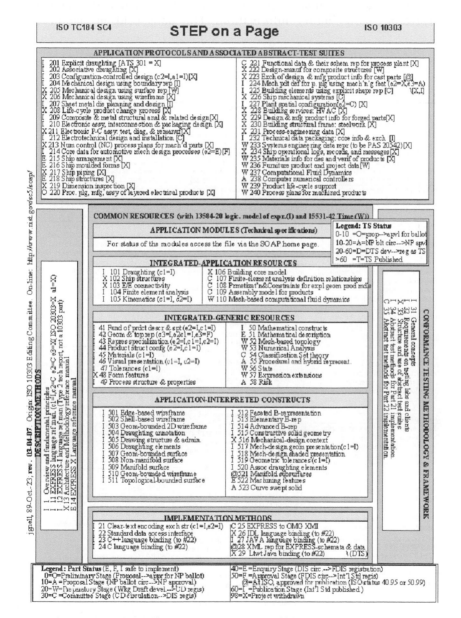

Figures 7.3 STEP on a page: components of the standard (schema) (SOAP).

7.3.2   ISO 13584 P-LIB

7.3.2.1   Purpose

ISO 13584 (13584–1) specifies the structure of a library system which provides an unambiguous representation and exchange of computer interpretable parts library information. The data held in the library

ISO TC184 SC4	STEP on a Page	ISO 10303

STEP on a Page provides a graphic summary of the progress of STEP, Standard for the Exchange of Product Model Data, the familiar name for ISO 10303. ISO TC184 SC4, Industrial-Automation Systems and Integration/Industrial Data develops the STEP standard.

Status of STEP Parts

Every part shown in the STEP on a Page has its status shown beside it. The status designators vary from "O" (the ISO preliminary stage) to "I" (International Standard--the stage in which the standard is published). Parts designated as "E, F" (levels of Draft International Standard) and "I" are considered advanced enough to allow software vendors to prepare implementations. The legend at the bottom of the page lists the corresponding ISO-project stage numbers next to the letter code.

Architecture of STEP

STEP on a Page attempts to show the STEP architecture by grouping the STEP parts into five main categories: description methods, implementation and conformance methodology, common resources, abstract-test suites, and application protocols.

Description Methods

From an architectural perspective, the description methods group forms the underpinning of the STEP standard. This includes part 1, Overview, which also contains definitions that are universal to the STEP. Also in that group, part 11, EXPRESS Language Reference Manual, describes the data-modeling language that is employed in STEP. Parts in the descriptive-methods group are numbered from 1 to 19.

Implementation & Conformance

The STEP implementation-methods group, the 20s series, describes the mapping from STEP formal specifications to a representation used to implement STEP.

The conformance-testing-

methodology-framework group, the 30s series, provides information on methods to test software-product conformance to the STEP standard, guidance for creating abstract-test suites, and the responsibilities of testing laboratories. The STEP standard is unique in that it places a very high emphasis on testing, and actually includes these methods in the standard itself.

Common Resources (IR, AIC, and AM)

At the next level is the common-resources group, the parts that contain the generic-STEP-data models. The common resources were formerly called integrated-information resources. These data models can be considered the building blocks of STEP, and they can help AP integration and interoperability because entities in the common-resources group are shareable across the application protocols that need them.

Categories of common resources are generic resources, application resources, and application-interpreted constructs, application modules, plus the Logical model of ISO 13584-20 and the Time model of ISO 15531-42. Integrated generic resources are generic entities that are used as needed by application protocols (AP below). Parts within generic resources have numbers between 40 and 60, and are used across the entire spectrum of STEP APs. The integrated-application resources contain entities that have slightly more context than the generic entities. The parts in the integrated-application resources are numbered in the 100s.

The 500 series are application-interpreted constructs, AICs. These are reusable groups of information-resource entities that make it easier to express identical semantics in more than one AP.

Application Modules are reusable groups of functional information requirements of applications that extend the AIC capability. The

functional groups, defined in enterprise-application terms, are aligned with groups of integrated-generic resources. The application modules comprise the 1000 series of parts, which are technical specifications that achieve consensus at the Committee stage. AMs offer an opportunity to represent functional capability in multiple APs with a lower standards-development cost.

Abstract-Test Suites (ATS)

The 300 series of parts, abstract-test suites, consists of test data and criteria that are used to assess the conformance of a STEP software product to the associated AP. SC4 requires that every AP or be associated with an abstract-test suite. The numbers assigned to ATSs exceed the AP numbers by exactly 100. Therefore, ATS 303 applies to AP203. On the graphic, the ATS status is shown in brackets [ ], following the AP name.

Application Protocols (AP)

At the top level of the STEP hierarchy are the more complex data models used to describe specific product-data applications. These parts are known as application protocols and describe not only what data is to be used in describing a product, but also how the data is to be used in the model. The APs use the integrated-information resources in well-defined combinations and configurations to represent a particular data model of some phase of product life. APs are numbered in the 200s. APs currently in use are the Explicit Draughting AP 201 and the Configuration Controlled Design AP 203.

ooOOoo
STEP on a Page was conceived and implemented by Jim Nell, National Institute of Standards and Technology. Updated 01-June-07

Figures 7.4   STEP on a page: components of the standard (details) (SOAP).

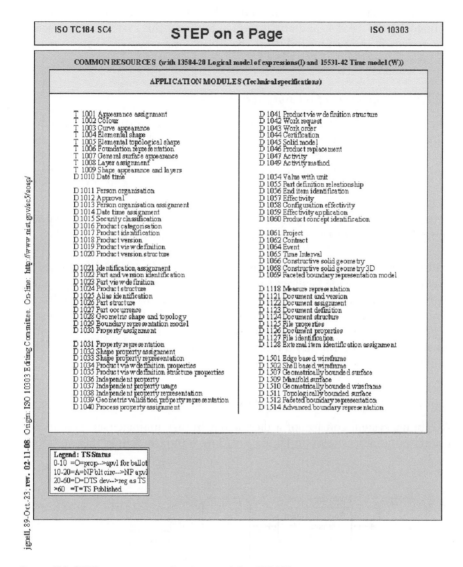

Figures 7.5 STEP on a page: application modules (SOAP).

are a description that enables the library system to generate various representations of the parts held in the library. The structure is independent of any particular computer system and permits any kind of part representation. The structure will enable consistent implementations to be made across multiple applications and systems.

ISO 13584 does not specify the content of a supplier library. The content of a supplier library is the responsibility of the library data supplier.

The library management system used in the implementation of the structure defined in ISO 13584, and any interface between this system and a user of the system is the responsibility of the library management system vendor and is not specified in ISO 13584.

7.3.2.2   Components of a library system

The components which form a neutral library system may be split into a number of functional areas which are illustrated in Figure 7.6.

User to computer system communication: The interface between the user and his computer system is not defined in ISO 13584. This would be application dependent and form part of the user interface supplied by a vendor as part of a computer system.

Interface to External Systems: The interface between a library system compliant with ISO 13584 and other software systems are:

a library Interrogation Interface: not defined in ISO 13584 but would be expected to provide facilities to select parts from the library and to define the orientation, position and representation category of the part selected;

a representation transmission interface, enabling the library system to send parts representations to the user computer system;

an input interface for library data, enabling the integration of supplier libraries within a library system.

7.3.2.3   Internal structure of a library system

A Library system consists of a dictionary, library management system and library content as shown in Figure 7.7. The standard defines these modules

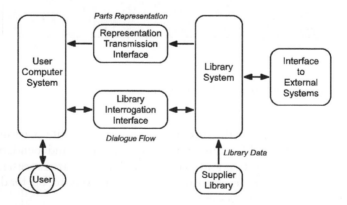

Figure 7.6 Functional areas of library usage.

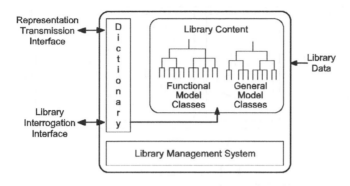

Figure 7.7 Library system.

by the requirements placed upon their functional behaviour. ISO 13584 does not standardise their implementation.

Dictionary   Consisting in a set of entries associated with a human-readable and computer-sensible representation of the meaning associated with each entry. The dictionary may be accessed by the user and referenced from library data. The Dictionary provides a referencing mechanism between library data obtained from different suppliers and enables the user to obtain an understandable view of the parts held in the library. The dictionary structure is specified in ISO 13584–42 (13584–42). A supplier library may contain only dictionary entries. These entries provide computer-referable identifiers for the concepts involved in some application domain.

Library Management System   Software system that enables the end user of the library to use the content of an integrated library and to load data into that library. The Library Management System is not standardised within ISO 13584.

Library content   Library data are structured into classes in accordance with the object oriented paradigm. Three kinds of classes are considered in ISO 13584. The contents of the three kinds of classes may be exchanged using the structure and exchange format specified in P-LIB.

General model classes enable library data suppliers to provide the definition of a collection of cognate parts considered as a part family. Functional model classes enable library data suppliers to provide various representations (e.g. geometric, schematics, procurement data etc.) for these collections of cognate parts. Functional view classes enable the specification of the kind of representation provided in the different functional model classes. Some functional view classes are standardised in the view exchange

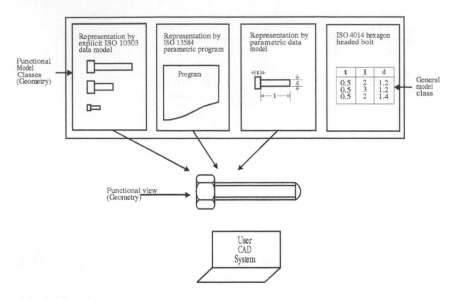

Figure 7.8 Structure of library contents.

protocol series of ISO 13584. A library data supplier may also provide the definition of their own functional view class. These three kinds of classes are illustrated in Figure 7.8.

When a library consists only of a dictionary, it only defines the concept associated with each class and with the properties of each class. When a library also contains a library content, this content defines the set of instances contained by each defined class.

When the user CAD system is compliant with an ISO 10303 STEP application protocol(s), the provisions contained in ISO 13584 ensure that it is possible from a library content to generate a functional view that is compliant with an ISO 10303 application protocol.

7.3.2.4   Fundamental principles of the standard

ISO 13584 separates the representation of information held in a parts library from the implementation methods used in data exchange. The standard makes use of a formal data specification language, EXPRESS, to specify information about the structure of a library. ISO 13584 separates information about the structure of a parts library from the information about different representations of each part or family of parts in the library. ISO 13584 permits information about part representation to be specified by different standards, and includes mechanisms which enable references to such descriptions (Pierra *et al.*, 1998).

7.3.2.5   Structure of the ISO 13584 series of parts

ISO 13584 is divided into series of parts, each with a unique function. Each series may have one or more parts. The series are listed below with their numbering scheme:

Conceptual description	Parts 10 to 19
Logical resources	Parts 20 to 29
Implementation resources	Parts 30 to 39
Description methodology	Parts 40 to 49
Conformance testing	Parts 50 to 59
View exchange protocol	Parts 101 to 199
Standardised content	Parts 500 to 599

Conceptual descriptions   They define the global conceptual framework and mechanisms developed to allow the portability of multi-supplier and multi-representation parts libraries, for exchanging and for updating. They present a problem domain analysis of the universe of discourse. They describe the concepts and choices made in the formulation of ISO 13584. The division of the whole task to be performed into a number of logical tasks that may be defined as a separate part of ISO 13584 is accomplished in the conceptual description series of parts.

Logical resources   The information model of parts library is provided by a set of resources. Each resource is comprised of a set of data descriptions in EXPRESS, known as resource constructs. One set may be dependent on other sets for its definition. Some resources constructs from ISO 10303 may be used to define ISO 13584 resources constructs. All the ISO 13584 resource constructs are defined in one part of the logical resources series. These resources may be used, but not modified, in a view exchange protocol.

Implementation resources   Each representation category may require a representation transmission interface to be implemented on a receiving CAD system to be able to interpret part models and to generate part views. The implementation resources specify the standardised representation transmission interfaces which may be referenced by a view exchange protocol. Each part of this series either specifies an interface, with the requirements for its implementation, or specifies the requirements for the implementation of one interface specified in other standards.

Description methodology   Providing rules and guidelines for library data suppliers, who may be standardisation organisations, part suppliers or functional model suppliers. These rules are intended to ensure consistency of a user Library. They are mandatory for the standardisation committees, in charge of specifying standardised dictionary data. They provide optional guidelines for part suppliers or functional model suppliers.

Conformance testing   Providing test cases and a set of requirements that any implementation shall meet before being accepted as conforming to this Standard.

View exchange protocol   Specifying one set of requirements for the exchange of one representation category of parts. Several view exchange protocols may refer to the same representation category. A view exchange protocol may introduce different options that may be selected by an implementation. The options are termed conformance classes. In this case the requirements of the view exchange protocol are specified separately for each conformance class.

Standardised content   It is intended to progressively define standardised dictionary entries which may be referenced by supplier libraries. This work will be done inside different standardisation committees following the methodology specified in the description methodology series of parts of ISO 13584. The parts of the standardised content specify the standardised dictionary entries corresponding to various application areas.

7.3.2.6   Use of library parts in product data

An ISO 13584 conforming exchange context provides for the exchange of library data intended to be stored in a user library. An ISO 10303 STEP conforming exchange context provides for the exchange of product data.

Three levels of interactions have been identified between these two levels of exchange.

Level 1   All information about a part generated in System A will be transferred to System B by means of ISO 10303 (see Figure 7.9).

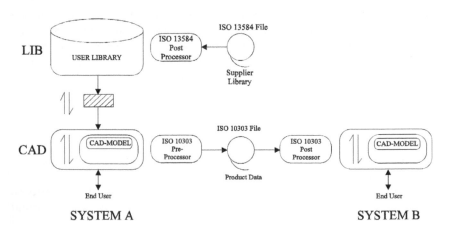

Figure 7.9 Libraries and product data exchange (level 1).

Level 2 Only that information is transferred from System A to System B, which is necessary to generate the same part from a Library 2 of the receiving System B at the required position and orientation. Library 1 and Library 2 both contain all the information about the part (see Figure 7.10).

Level 3 That information is transferred from System A to System B which is necessary to generate the same part information on the receiving System B without any assumption about the content of Library 2. This means that the transferred data also contains a subset of Library 1 (see Figure 7.11).

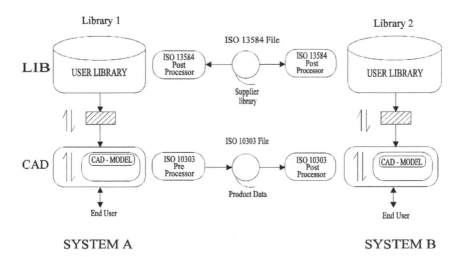

Figure 7.10 Libraries and product data exchange (level 2).

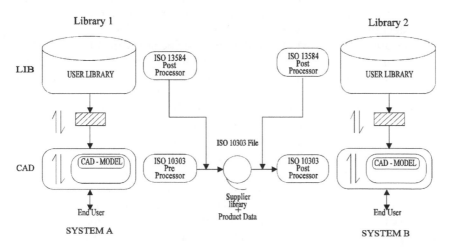

Figure 7.11 Libraries and product data exchange (level 3).

The information models specified in ISO 13584 are intended to enable these three levels of interaction.

7.3.3   ISO 18629: Process Specification Language: (PSL)

ISO 18629 is the newest in the family of standards aimed at facilitating interoperability for industrial data integration (of products and processes) in industrial applications in TC 184. Standardised within a joint committee, ISO TC 184 SC4/SC5, PSL provides a generic language for process specifications applicable to a broad range of specific process representations in manufacturing and other applications. PSL is an ontology for discrete processes written in the Knowledge Interchange Format (KIF) (Genesereth and Fikes, 1992) itself an ISO candidate in (ISO/JTC1, 1999), (Common Logic, 2004). Each concept in the PSL ontology is specified with a set of definitions, relations and axioms all formally expressed in KIF. Relations specify types of links between definitions or elements of definitions; axioms constrain the use of these elements. In addition, the PSL ontology is based on set theory, first order logic, and situation calculus (Etchemendy, 1992). Because of this reliance on theories, every element in the PSL language can be proven for consistency and completeness (Gruninger, 2003). At the time of this writing, approximately half of the PSL definitions, relations and axioms have been proven to be consistent with the base theories.

PSL is an international standard for providing semantics to the computer-interpretable exchange of information related to manufacturing and other discrete processes. Taken together, all the parts contained in PSL provide a language for describing processes throughout the entire production within the same industrial company or across several industrial sectors or companies, independently from any particular representation model. The nature of this language makes it suitable for sharing process information during all the stages of production. The process representations used by engineering and business software applications are influenced by the specific needs and objectives of the applications. The use of these representative models vary from one application to another, and are often implicit in the implementation of a particular application. One of the manufacturing models on which the PSL ontology is built is provided by the information models of the ISO 15531 MANDATE standard (standardisation of manufacturing management information) (Cutting-Decelle et al., 2000–2001), particularly for resource management.

A major purpose of PSL is to enable the interoperability of processes between software applications that utilise different process models and process representations. As a result of implementing process interoperability, economies of scale are made in the integration of manufacturing applications.

All parts in ISO 18629 are independent of any specific process representation or model used in a given application. Collectively, they provide a structural framework for interoperability. PSL describes what elements should constitute interoperable systems, but not how a specific application implements these elements. The purpose is not to enforce uniformity in process representations. As objectives and design of software applications vary the implementation of interoperability in a application must necessarily be influenced by the particular objectives and processes of each specific application.

7.3.3.1 Architecture and content of ISO 18629

PSL (ISO 18629–1, 2004) is organised in a series of parts using a numbering system consistent with that adopted for the other standards developed within ISO TC184/SC4. PSL contains Core theories (Parts 1×), External Mappings (Parts 2×) and definitional extensions (Parts 4×). This discussion focuses on Parts 1× and 4×; these parts contain the bulk of ISO 18629, including formal theories and the extensions that model concepts found in applications. Parts 1× are the foundation of the ontology, Parts 4× contain the concepts useful for modeling applications and their implementation. Table 7.1 presents the organisation of ISO 18629. Except noted otherwise, PSL version 2.2 is presented.

Core theories (Parts 1×) Core Theories include the PSL-Core, the Outer Core, Duration and Ordering theories, Resource theories and Actor and Agent theories. The core theories are contained in the parts 1× and based on first-order logic. They model basic entities necessary for building the PSL extensions. The PSL-Core and Core theories pose primitive concepts (those with no definition), function symbols, individual constants, and a set of axioms written in the language of PSL. Table 7.2

Table 7.1 Organisation of ISO 18629

Series	Number	Name
Core theories	ISO IS 18629–1	Overview and basic principles
	ISO IS 18629–11	PSL-Core
	ISO IS 18629–12	Outer Core
	ISO CD 18629–13	Duration and ordering theories
	ISO CD 18629–14	Resource theories
	ISO WD 18629–15	Actor and agent theories
External Mappings	ISO 18629–2x	Mappings to EXPRESS, UML, XML
Definitional extensions	ISO DIS 18629–41	Activity extensions
	ISO DIS 18629–42	Temporal and state extensions
	ISO CD 18629–43	Activity ordering and duration extensions
	ISO CD 18629–44	Resource extensions
	ISO WD 18629–45	Process intent extensions

illustrates the primitives found in the PSL-Core. These primitives and all the definitions in PSL are written in KIF for computer interoperability but the KIF writing is not shown here for the sake of readability. For KIF sentences expressing these relations and functions the reader is referred to the PSL Web site.

Core theories are required to formally prove that extensions are consistent with each other, and with the core theories. The core theories are at the root of the PSL ontology against which every item that claims to be PSL compliant must be tested for consistency. They are a unique feature of PSL as no other standard in SC4 lends itself to formal, logic-based

Table 7.2 Concepts in PSL-Core (ISO IS 18629–11)

PSL-Core primitives	Type	Informal definitions and axioms
Activity	Relation	Everything is either an activity, an activity occurrence, a timepoint or an object. Objects, activities, activity occurrences and timepoints are all distinct kinds of things (disjoint classes)
Activity_occurrence	Relation	An activity occurrence is associated with a unique activity. But there are activities without occurrences
Timepoint	Relation	Given any timepoint t other than inf−, there is a timepoint between inf− and t. Given any timepoint t other than inf+, there is a timepoint between t and inf+
Object	Relation	An object participates in an activity at a given timepoint and only at those timepoints when both the object exists and the activity is occurring
Before	Relation	The before relation only holds between timepoints. It is a total ordering, irreflexive and transitive relation
Occurrence_of	Relation	Every activity occurrence is the occurrence of some activity and associated with a unique activity
Participates_in	Relation	The participates_in relation only holds between objects, activities and timepoints, respectively
Beginof	Function	The beginning of an activity occurrence or of an object are timepoints
Endof	Function	The ending of an activity occurrence or of an object are timepoints
inf+	Constant	Every other timepoint is before inf+
inf−	Constant	The timepoint inf− is before all other timepoints

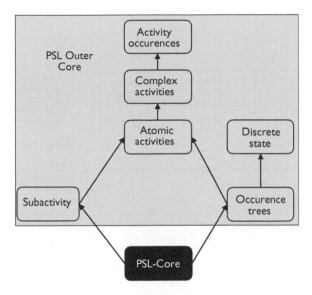

Figure 7.12 PSL Outer Core definitions and their dependencies (ISO IS 18629–12).

proof. Figure 7.12 illustrates concepts in the PSL Outer Core and their dependencies (ISO IS 18629–1, 2004).

Figure 7.13 extends Figure 7.12 to focuses on Duration, Ordering and Resource Requirements theories.

Domain-specific definitional extensions (Parts 4×)   The extensions to the Core and Outer Core are used to represent the actual processes in an application. All terms in the extensions are given definitions using the set of primitive concepts axiomatised in the core theories. This ensures that definitions are consistent with PSL. A software application will typically use the concepts defined in the extensions, rather than the concepts in the Core and Outer Core, which are necessary to define the extensions but have little expressivity.

In Figure 7.14, a definitional extension (Parts 4×) is represented as a slice of the pie. It specifies concepts and definitions for all kinds of (practical) concepts and are written using the Core, Outer Core and theories. Some definitional extensions also use concepts defined in other extensions. Figure 7.14 shows that a concept belonging to an extension (blue triangle) is specified using concepts of the Core Outer Core and another extension. But the Core and Outer Core alone are not sufficient to represent meaningfully an application's semantics for the purpose of interoperability.

Table 7.3 gives examples of definitional extensions, and the core theories each extension relies upon. It is to be noted that the organisation of the Extensions into Activity Extensions, Temporal Extensions, etc. is here for

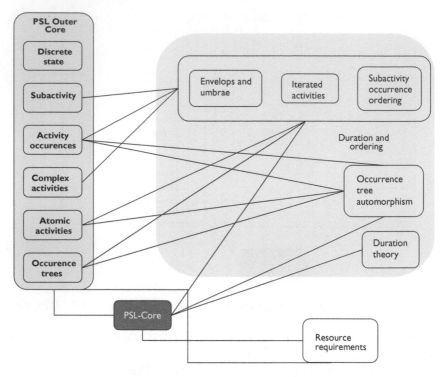

Figure 7.13 Core and Outer Core dependencies for duration and ordering and resource requirements theories (ISO CD 18629–13), (ISO CD 18629–14).

readability and ease of use of the standard. The organisation itself does not affect the concepts in PSL: for instance a concept may be moved from one extension to another without affecting the PSL ontology or the concepts defined in the extension. In other words, to be a valid part of the PSL ontology extensions do not need to belong to one or another of the categories in the left column. However, each concept must conform to the Core Theories in the middle column.

7.3.3.2   Interoperability with PSL and conformance to the standard

The main purpose of PSL is to establish a computer language for exchanging processes between software applications such as CAD, and project design software. As a specification language, PSL can be considered as a specification tool of the information and knowledge related to manufacturing management, as modelled by the MANDATE standard (ISO IS 15531–1, 2002).

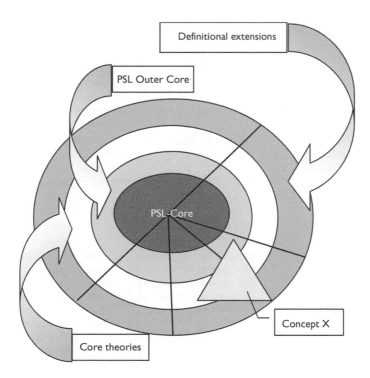

Figure 7.14 Architecture of the PSL ontology.

7.3.3.2.1 THE CHALLENGES OF INTEROPERABILITY

The obstacles to interoperability of data regarding syntax of two applications are common, and usually dealt with parsers. Obstacles due to semantic problems, that is problems about the 'meaning' of a software object or entity are less visible. Lack of semantic reconciliation may introduce errors even if syntax mapping is correct. Without a standard like PSL, the semantic mapping may be performed in an ad hoc manner by a developer.

Figure 7.15 presents an example from the transportation industry, where a truck is represented as a vehicle, a mobile resource, or a truck. Delivery mechanisms not represented here may also include transportation for some applications. If there is no interoperability of processes, applications that use this terminology may be incompatible. This leads to re-inputting entries manually in the application chain. In the example in Figure 7.16, Material designates two different things: a Resource and a Work in Progress and a Resource. Resource a Material, a Machine-tool and a Stock.

Syntactic interoperability does not resolve these conflicts, and decisions as to which concept in Application A matches a concept in Application B is

Table 7.3 Examples of PSL concepts defined in extensions

Definitional extensions (Parts 4x)	Core theories depended upon	Some examples of definitions
Activity Extensions (Part 41) (ISO DIS 18629–41, 2004)	Complex Activities	Deterministic and non-deterministic activities Concurrent activities Spectrum of activities
Temporal and State Extensions (Part 42) (ISO DIS 18629–42, 2004)	Complex Activities, Discrete States	Preconditions, effects conditional activities triggered activities
Activity Ordering and Duration Extensions (Part 43) (ISO CD 18629–43, 2004)	Sub-activity occurrence ordering, iterated occurrence ordering, duration	Complex sequences and branching Iterated activities Duration-based constraints
Resource Extensions (Part 44) (ISO CD 18629–44, 2004)	Resource Requirements Resource set theory Sub-activity Occurrence Ordering Resource Requirements	Reusable, consumable, renewable, and deteriorating resources, substitutable resources resource pools, Resource paths Processor activities

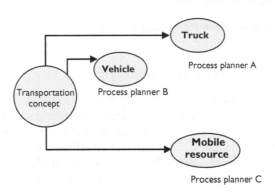

Figure 7.15 Incompatible content representation.

left to the developer of parsers. The benefit of PSL is to formally encode each application's concept or vocabulary in a rigorous representation language. When two applications sharing data are expressed in PSL, the conflicts and semantic gaps are highlighted and a resolution is proposed. In essence, expressing the concepts of an application with PSL produces a detailed analysis of processes, and on this basis two applications can be reconciled.

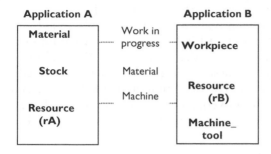

Figure 7.16 Semantic conflict for resource.

7.3.3.2.2   INTEROPERABILITY AND CONFORMANCE

From the point of view of ISO 18629 two applications can inter-operate if they are conformant with the same set of ISO 18629 extensions (ISO IS 18629–1). Software applications that claim conformance to PSL will:

specify processes from their application into the KIF language. This is the set of terms used by the application that refer either to processes in the application or relations among these processes;
provide translation definitions between their processes represented in KIF and PSL definitions;
implement syntactic translators between their applications and PSL process descriptions.

Another requirements not discussed in this chapter is that there exists a grammar using the same representation as PSL grammar for the application processes, using the Backur Naus form.

In practice, two applications do not exchange data about all their processes in one exchange. Only one or a set of processes at one time will exchange data. After identifying the concept to be exchanged, the steps outlined in the standard can be followed as:

the processes are defined and expressed using KIF syntax;
the concepts contained in the processes (their names, relationship to other processes, conditional expressions) are further defined. In other words, the application's entities are given KIF definitions;
a translation is provided between the application's entities definition and PSL definitions.

At this point in the procedure, Applications A and B's processes have been expressed using PSL terms and KIF syntax. Each has a one-to-one correspondence between their process definition and a PSL definition. On this basis, data for the relevant process can be exchanged.

Following this procedure does not allow a software application to claim conformance to PSL according to ISO 18629, but it is sufficient for process exchange with another application. To this purpose, the National Institute of Standards has implemented a 'question wizard' (PSL, Wizard) to facilitate the expression of any process with PSL definitions and in KIF syntax. A user specifies a process in details by answering questions and checking boxes for their process. The wizard returns a definition for the process using PSL.

7.3.3.2.3   USER DEFINED EXTENSIONS

User defined extensions of PSL are extensions that introduce new primitive concepts. Typically, current extensions are sufficiently rich to express processes in existing software applications. However, the case where an application concept is not represented may arise. In this case, PSL can be extended to include a new extension by expressing it using the PSL Core, Outer Core and definitions in existing extensions. The axioms in any extension that introduces new primitives must be consistent with the axioms of PSL-Core. User-defined extensions are needed when PSL is applied to domains that have not been yet dealt with in the extensions.

Research work has been done or is currently on-going, showing examples or interoperability among software tools using PSL, notably at the University of Stanford (CIFE) (Law, 2001) (Cheng *et al.*, 2003), and at the University of Loughborough (Cutting-Decelle *et al.*, 2000) (Cutting-Decelle *et al.*, 2002) (Cutting-Decelle *et al.*, 2004) (Tesfagaber *et al.*, 2002). We present below an example of process exchange in construction using PSL.

7.3.3.2.4   EXAMPLE OF PROCESS EXCHANGE IN CONSTRUCTION USING PSL

Scenario   To illustrate interoperability for the construction domain, the following scenario has been designed:

The design and construction of an office building includes an exchange of information and data with the purpose of fitting a metal door to a metal wall frame. Conception, estimation of costs and project planning must be studied for this scenario. Software applications used for this study include a design application using AutoCAD, a cost simulation software and a project planner using MS-Project for the planning phase. A related scenario would be the exchange of processes for integrating a new supplier. Figure 7.17 illustrates this scenario. The process exchange is described in details for the AutoCAD and MS Project applications (Tesfagaber, 2004).

Process interchange with PSL   First the Architectural Design File is written using the syntax and terminology of the AutoCAD application.

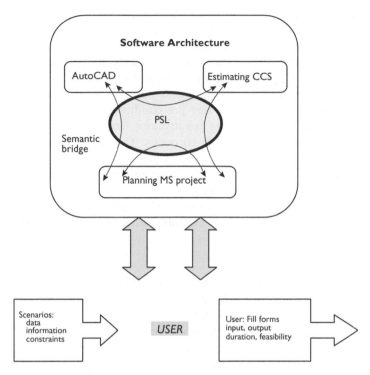

Figure 7.17 Data and information exchange scenario.

Source: Tesfagaber (2004).

Second this file must be parsed to a KIF file, still using AutoCAD's terms and relations. This syntax translation may be executed with a parser between KIF and the AutoCAD's syntax. Thus an ontology for the design application is built. Third, the AutoCAD application's ontology is semantically translated into PSL terminology. This step is usually done manually or using the 20-question wizard developed at NIST for that purpose. This involves in-depth understanding of the processes in the AutoCAD application and may require consulting the documentation. It also introduces as many constraints and relations as possible on the AutoCAD terminology. Constraints and relations are taken from PSL and necessary for specifying in details what the AutoCAD terminology means. The result is a file where the AutoCAD is expressed using PSL concepts under specific conditions.

In parallel, the same process is performed for the process planning application using MS-Project terminology. Once both applications have been expressed in KIF and specified using PSL concepts, a inverse file containing constrained PSL concepts equivalent to the MS-Project application is created. Using these two files (AutoCAD using PSL concepts and PSL concepts corresponding to MS-Project) process data can be exchanged under explicit conditions.

Below is the PSL translation of a process named door-assembly in AutoCAD developed using the 20-question wizard (Figure 7.18). This process is an activity. Its initiation depends on the state of other activities prior to this one (markov pre-conditions), but not on time or duration allocated to the activity. For instance, 'make door frame' may be required for door frame assembly to occur. The result of the process is affected by the initial conditions existing prior to the process but not the duration. All occurrences of the 'door assembly' activity have the same effect and are also time-independent.

The specification of an MS task using PSL is given below. This specification intends to verify if a door-assembly process in AutoCAD can be

```
(forall (?a)
      (iff(doorframe+assembly ?a)
         (and (activity ?a)
                (constrained ?a)
                (markov_precond ?a)
                (rigid_time ?a)
                (rigid_mixed ?a)
                (context_free ?a)
                (markov_effects ?a)
                (nontemporal ?a)
                (rigid_mixed_effects ?a))))
```

Figure 7.18 Door-assembly process described with PSL.

```
(forall (?a)
      (iff(task ?a)
         (and (activity ?a)
                (constrained ?a)
                (markov_precond ?a)
                (time_precond ?a)
                (mixed_precond ?a)
                (context_free ?a)
                (rigid_state_effects ?a)
                (rigid_time_effects ?a)
                (rigid_mixed_effects ?a))))
```

Figure 7.19 The MS-task described by PSL.

equated to a task in MS-project. If it was, only a syntactic parser for values of variables between AutoCAD and MS-project is necessary in information exchange. If not, the development of translation software may be assisted by providing an in-depth analysis of semantics and resolve the discrepancies using PSL.

In this example (Figure 7.19), a task is a constrained activity that cannot occur unless other activities have previously occurred: it depends on the state of another activity. This is similar to AutoCAD. However, an MS task may also be bound by starting time, duration of the activity, or a combination of these with a state pre-condition. Therefore, if a parser is designed to translate from the AutoCAD file for door-assembly to a task in MS project, the parser must take into account the fact that time does not exist for the AutoCAD process so that the AutoCAD process is not equivalent to the MS task. In other words the parser can only partially input a new task based on the information provided by AutoCAD. Semantic encoding using PSL has here highlighted a potential source of error for automatic translators between two applications. In this case, it is determined that the AutoCAD process of door-assembly can be translated to an MS task constrained by the occurrence of activities, but additional constraints regarding time also exist. One possible solution to this obstacle is to enter by hand the values for time constraints and duration.

User defined extensions   User defined extensions of PSL are extensions that introduce new primitive concepts, for instance for a domain where PSL has not been used before. Typically, current extensions and existing processes are sufficiently rich in PSL to express existing software applications. However, the case where an application concept is not represented may arise. In this case, PSL can be extended to include a new concept or extension by expressing it using the PSL Core, Outer Core and definitions in existing extensions. The axioms in any extension that introduces new primitives must be consistent with the axioms of PSL-Core.

7.4 De facto standard developed by the International Alliance for Interoperability (IAI) Industry Foundation Classes (IFCs)

7.4.1 The IAI community

The IAI is an international consortium of regional chapters registered and listed as non-for-profit organizations in North America, United Kingdom, Germany, France, Scandinavia, Japan, Singapore, Korea and Australia. Currently the IAI has about 650 membership organizations world-wide, being construction companies, engineering firms, building owners and operators, software companies and academic institutions. The vision of the IAI is: 'to provide a universal basis for process improvement and information sharing in the construction and facilities management industries' (IAI, 2001).

The vision is supported by the IAI mission statement: 'to define, promote and publish the Industry Foundation Classes (IFC), a specification for sharing data throughout the project life-cycle, globally, across disciplines and across technical applications'.

More information about the IAI is available at: http://www. iai-international.org.

7.4.2   The IFCs

The IFC are a data sharing specification, written in EXPRESS (10303–11, 1994), the dedicated formal language developed within the ISO 10303 STEP standard. Content according to IFC is currently exchanged between IFC compliant software applications using the Clear text encoding of the exchange structure, the STEP physical file (10303–21, 1994).

The scope of the IFC specification is the project life-cycle of construction facilities, including all phases as identified by generic process protocols for the construction and facilities management industries, such as: Demonstrating the need, Conception of need, Outline feasibility, Substantive feasibility study and outline financial authority, Outline conceptual design, Full conceptual design, Co-ordinated design, procurement and full financial authority, Production information, Construction, Operation and maintenance.

Development of IFC is guided by versions and releases, which do extend the scope successively. The processes supported by the current IFC2x specifications are: Outline conceptual design, Full conceptual design, Co-ordinated design, procurement and full financial authority, Production information, Construction, Operation and maintenance.

The target applications to exchange and share information according to IFC2x are: CAD Systems, HVAC design systems, Electrical design systems, Formwork design and scheduling systems, Structural analysis systems, Energy simulation systems, Quantity take-off systems, Cost estimation systems, Production scheduling systems, Clash-detection systems, Product information providers, Steel and Timber frame construction systems, Prefab systems, stand-alone visualisation tools and others.

7.4.3   IFC Model Architecture

7.4.3.1   Architecture principles

The IFC Model Architecture has been developed using a set of principles governing its organisation and structure. These principles focus on basic requirements and can be summarised as (IAI, 2000):

To provide a modular structure to the model.
To provide a framework for sharing information between different disciplines within the AEC/FM industry.

To ease the continued maintenance and development of the model.
To enable information modelers to reuse model components.
To enable software authors to reuse software components.
To facilitate the provision of better upward compatibility between model releases.

The IFC Model Architecture provides a modular structure for the development of model components, the 'model schemata'. There are four conceptual layers within the architecture, which use a strict referencing principle. Within each conceptual layer a set of model schemata are defined.

1   The first conceptual layer provides Resource classes used by classes in the higher levels.
2   The second conceptual layer provides a Core project model. This Core contains the Kernel and several Core Extensions.
3   The third conceptual layer provides a set of modules defining concepts or objects common across multiple application types or AEC industry domains. This is the Interoperability layer.
4   Finally, the fourth and highest layer in the IFC Model is the Domain layer. It provides set of modules tailored for specific AEC industry domain or application type.

The architecture operates on a 'gravity principle'. At any layer, a class may reference a class at the same or lower layer but may not reference a class from a higher layer. References within the same layer must be designed very carefully in order to maintain modularity in the model design. Inter-domain references at the Domain Models layer must be resolved through 'common concepts' defined in the Interoperability layer. If possible, references between modules at the Resource layer should be avoided in order to support the goal that each resource module is self-contained. However, there are some low level, general purpose resources, such as measurement and identification that are referenced by many other resources.

7.4.3.2   Gravity principle: see Figure 7.20

1   Resource classes may only reference or use other Resources.
2   Core classes may reference other Core classes (subject to the limitations listed in 3) and may reference classes within the Resource layer without limitations. Core classes may not reference or use classes within the Interoperability or Domain layers.
3   Within the Core layer the 'gravity principle' also applies. Therefore, Kernel classes can be referenced or used by classes in the Core Extensions but the reverse is not allowed. Kernel classes may not reference Core Extension classes.

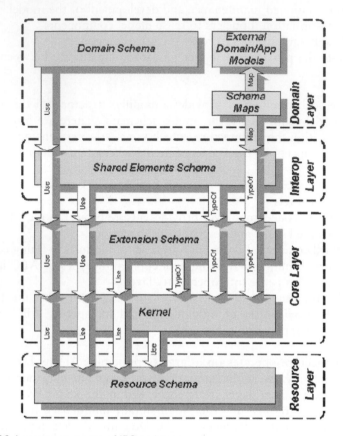

Figure 7.20 Layering concepts of IFC architecture.

Source: IAI (2000).

4   Interoperability layer classes can reference classes in the Core or
    Resource layers, but not in the Domain layer.
5   Domain layer classes may reference any class in the Interoperability,
    Core and Resource layers.

7.4.4   IFC model architecture decomposition

The IFC model architecture for IFC 2 × consists of the following layers:

Resource layer
Core layer
Kernel
Extensions
Interoperability layer
Domain layer

7.4.4.1 Resource layer

Resources form the lowest layer in IFC Model Architecture and can be used or referenced by classes in the other layers. Resources can be characterised as general purpose or low level concepts or objects that are independent of application or domain need (i.e. they are generally rather than specifically useful) but which rely on other classes in the model for their existence. For instance, geometry is a widely used resource whose specification is independent of domain. However, an object within a domain must be defined before its geometry can exist.

Exceptions to this characterisation include classes from the Utility and Measure Resources that are used by other, higher-level resource classes.

7.4.4.2 Core layer

The Core forms the next layer in IFC Model Architecture. Classes defined here can be referenced and specialised by all classes in the Interoperability and Domain layers. The Core layer provides the basic structure of the IFC object model and defines most general concepts that will be specialized by higher layers of the IFC object model (Figure 7.21).

The Core includes two levels of generalisation:

1 *The Kernel* Provides all the basic concepts required for IFC models within the scope of the current IFC Release. It also determines the model structure and decomposition. Concepts defined within the kernel are, necessarily, generalised to a high level. It also includes fundamental concepts concerning the provision of objects, relationships, type definitions, attributes and roles. The Kernel can be seen as a template model that defines the form in which all other schema within the model are developed (including all extension models). Its constructs are very general and are not AEC/FM specific, although they will only be used for AEC/FM purposes due to the specialization by Core Extensions. The Kernel constructs are a mandatory

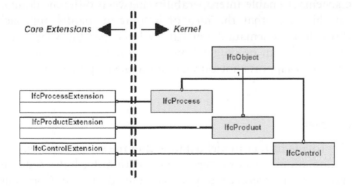

Figure 7.21 Core extensions from kernel classes.

Source: IAI (2000).

part of all IFC implementations. The Kernel is the foundation of the Core Model. Kernel classes may reference classes in the Resource layer but may not reference those in the other parts of the Core or in higher-level model layers.

2 *Core Extensions* Provide extension or specialisation of concepts defined in the Kernel. They are the first refinement layer for abstract Kernel constructs. More specifically, they extend those constructs for use within the AEC/FM industry. Each Core Extension is a specialisation of classes defined in the Kernel and develops further specialisation of classes rooted in the IFCKernel. Additionally, primary relationships and roles are also defined within the Core Extensions. A class defined within a Core Extension may be used or referenced by classes defined in the Interoperability or Domain layers, but not by a class within the Kernel or in the Resource layer. References between Core Extensions have to be defined very carefully in a way that allows the selection of a singular Core Extension without destroying data integrity by invalid external references.

Goals for Core layer design are:

• definition of those concepts that are common to all parts of the model and that later can be refined and used by various interoperability and domain models.
• pre-harmonisation of domain models by providing the set of common concepts.
• stable definition of the object model foundation to support upgrade compatible IFC Releases.

7.4.4.3   Interoperability layer

The main goal in the design of Interoperability Layer is the provision of schemata that define concepts (or classes) common to two or more domain models. These schemata enable interoperability between different domain models. It is at this layer that the idea of a 'plug-in' model approach emerges. It is through the schemata defined at the Interoperability layer that multiple domain models can be plugged into the common IFC Core. The 'plug-in' approach also supports outsourcing of the development of domain models.

7.4.4.4   Domain layer

Domain Models provide further model detail within the scope requirements for an AEC/FM domain process or a type of application. Each is a separate model that may use or reference any class defined in the Core and Independent Resource layers. Examples of Domain Models are Architecture,

HVAC, FM, Structural Engineering etc. An important purpose of Domain Models is to provide the 'leaf node' classes that enable information from external property sets to be attached appropriately.

7.4.5   Connecting external models to the IFC Model

Fully harmonised IFC Domain Models are directly connected the Core definitions. Domain Models that are not fully harmonised have to provide appropriate connection to relevant IFC class definitions in order to use the IFC model framework. Such models may be developed according to different technical architectures and methodologies but might need to be used in conjunction with the IFC model at some point.

The means of achieving this is through the use of a connection mechanism. The main requirements for connection are the facilitation of:

1   Connection of externally developed, non harmonised, Domain Models via a connection that provides a mapping mechanism down to Core and Interoperability definitions. The definition of the connection is in the responsibility of the Domain Model developer and is part of the Domain Model Layer.
2   Establish an inter-domain exchange mechanism above the Core to enable interoperability across domains. This includes a container mechanism to package information. Therefore a connection is used where the definition of the connection is the responsibility of all Domain Models that share its use.

Connections are based on Core Extension definitions and enhance those Core Extension definitions. Those enhancements provide common concepts for all Domain models that might further refine these concepts. As an example, the Building Element provides the definition of a common wall, whereas the Architectural Domain Model will enhance this common wall with its private subtypes and type definitions. A connection that is used by several Domain Models therefore provides a level of interoperability through shared connection definitions.

Non-IFC harmonised models can be connected to the IFC Core Model through a specifically defined mapping. For specific high-level inter-domain exchange that cannot be satisfied by common definitions in the Core, connection through mapping may provide a specific inter-domain exchange capability.

7.4.6   Overall architecture

The following diagram (Figure 7.22) shows the complete set of IFC 2× model schema organised according to the layer at which they exist.

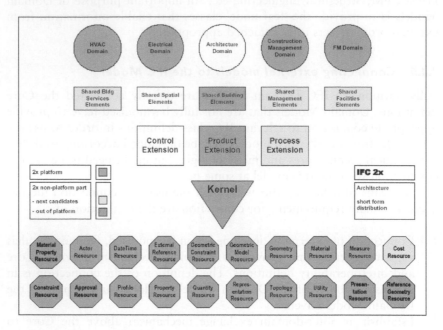

Figure 7.22 IFC 2× overall architecture.

Source: IAI (2000).

Note that all schema are named in a manner that enables identification of their architecture layer:

- Schema at the resource layer are suffixed with the term 'Resource'.
- Schema at the core extension layer are suffixed with the term 'Extension' (other than the Kernel schema which is considered to be a special case).
- Schema at the interoperability layer are suffixed with the term 'Elements'.
- Schema at the domain layer are suffixed with the term 'Domain'.

IFC 2× have been endorsed by the ISO organisation as the ISO/PAS 16793 in November 2002 (IAI, 2001).

7.5   Conclusion: contribution of different standards to interoperability in construction

We have shown to what extent standards-based approaches can be helpful to facilitate information sharing and interoperability among software applications commonly used in manufacturing, and in manufacturing management. Most of the time, technical terms handled by those applications look similar or, even worse, are exactly the same – however their meaning is different.

In the four standards described above, technical terms are established more or less on the same 'construction-flavoured' vocabulary, but are very different, with multiple interpretations of the same terms in each standard. Given its properties, and its structure, ISO 18629 PSL can be considered as a powerful interoperability 'tool' for the information systems of the enterprises. It introduces economy of scales – each application only needs to provide interoperability to PSL once for information exchange (Figures 7.23 and 7.24). If an application changes, it is up to the developers of this application to provide new translations to PSL. Thus only one application in the chain of inter-operation is affected and not the others.

However, implementation of standards is non-trivial and costly. It is not until the use and implementation of these standards in a particular industry has reached 'critical mass' that costs will decrease. STEP already has made

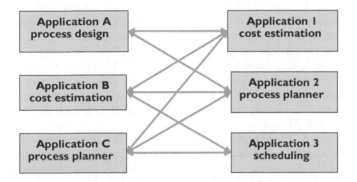

Figure 7.23 Information exchange without PSL.

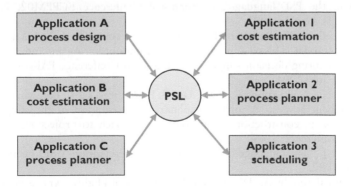

Figure 7.24 Information exchange with PSL.

great strides in this direction. The construction industry may benefit from the lessons learned in other domains, particularly for Concurrent engineering based approaches of construction projects (see Chapter 4).

7.6   Acknowledgement

The authors wish to thank G. Tesfagaber, PhD student and Prof. N. M. Bouchlaghem, Supervisor of her research work at the University of Loughborough.

7.7   References

Anumba C. J., Cutting-Decelle A. F., Baldwin A. N., Dufau J., Mommessin M. and Bouchlaghem N. M., 'Introduction of Concurrent Engineering concepts into an integrated Product and Process Model', Concurrent Engineering in Construction Conference, 1999.

Cheng J., Gruninger M., Sriram R. D. and Law, K. H., 'Process specification language for project scheduling information exchange', *International Journal of IT in AEC*, Vol. 1, Issue 4, December 2003.

Cutting-Decelle A. F. and Dubois J. E., 'Heterogeneous information exchanges in the construction industry – the SHYDOX/MATCOMP/Xi prototype', in *Heterogeneous Information Exchange and Organizational Hubs*, Bestougeff, Dubois and Thuraisingham eds, Kluwer Academic Publishers, 2002.

Cutting-Decelle A. F., Michel J. J. and Schlenoff C., 'Manufacturing and construction common process representation: the PSL approach', CE00 Lyon, 2000.

Cutting-Decelle A. F., Michel J. J. and Schlenoff C., 'Integration of industrial management information and interoperability: MANDATE + PSL a combined approach', MICAD 2000.

Cutting-Decelle A. F., Anumba C. J., Baldwin A. N., Bouchlaghem N. M. and Tesfagaber G., 'PSL: a common language for the representation and exchange of process information in construction', 1st International Conference on Innovation in AEC, CICE, University of Loughborough, UK, July 2001.

Cutting-Decelle A. F., Anumba C. J., Baldwin, A. N., Bouchlaghem N. M. and Tesfagaber G., 'Exchanges of process information between software tools in construction: the PSL language', International Conference ECPPM02, Portoroz, Slovenia, September 2002.

Cutting-Decelle A. F., Pouchard L. C., Das B. P., Young R. I. and Michel J. J., 'Utilising standards based approaches to information sharing and interoperability in manufacturing decision support', International Conference FAIM04, Toronto (Canada), July 2004.

Cutting-Decelle A. F., Young R. I. M., Das B. P., Anumba C. J., Baldwin A. N. and Bouchlaghem N. M., 'A multi-disciplinary representation of the supply chain information in construction: an innvoative approach to project management', TMCE Conference, 2004.

Etchemendy J., 'The language of first-order logic', 3rd edition, CSLI Lecture Notes No. 34, 1992.

Fikes R. and Farquahr A. 'Distributed Repositories of Highly Expressive Reusable Ontologies'. *IEEE Intelligent Systems and their Applications 14, no.2 (March–April, 1999)*.

FOLDOC, Free On-Line Dictionary of Computing, Imperial College Department of Computing, http://wombat.doc.ic.ac.uk/foldoc/

Genesereth M. and Fikes R. E., 'Knowledge Interchange Format, Version 3.0. Reference Manual', Technical Report Logic-92-1. Computer Science Department, Stanford University, Stanford, CA (January), 1992.

Gomez-Perez A., 'Knowledge Sharing and Re-Use', In *The Handbook of Applied Expert* Systems, J. Liebowitz, ed. Boca Raton, 10:1–10:36, 1998.

Gruber T., 'A Translation Approach to Portable Ontology Specifications', Knowledge Acquisitions 5 (May): 199–220, 1993.

Gruninger M., 'A guide to the ontology of the Process Specification Language', in *Hand-book on Ontologies in Information Systems*, R. Studer and S. Staab eds, Springer-Verlag, 2003.

http://www.mel.nist.gov/psl (PSL Web site).

http://www.iso.ch (ISO Web site).

http://www.tc184-sc4.org (ISO TC 184 / SC4 Web site).

http://philebus.tamu.edu/cl/ (Common Logic Web site).

http://www.whatis.com

http://www.ariadne.ac.uk/issue24/interoperability/intro.html, 2000

http://www.nist.gov/sc5/soap/

Hunhs M. N. and Singh M. P., 'Ontologies for Agents.' *IEEE-Internet Computing* 1, no. 6 (November–December), 1997.

IAI, IFC release 2x, IFC technical Guide, 2000.

IAI, Transposing the IFC2x Specification as ISO Standard, 2001.

ISO 10303–1, 'Industrial automation systems and integration – Product data representation and exchange – Part 1: overview and fundamental principles', 1994.

ISO 10303–11, 'Industrial automation systems and integration – Product data representation and exchange – Part 11: the EXPRESS language reference manual, 2003.

ISO 10303–21, 'Industrial automation systems and integration – product data representation and exchange – Part 21: Implementation methods: Clear text encoding of the exchange structure', 1994.

ISO 10303–22, 'Industrial automation systems and integration – Product data representation and exchange – Part 21: Standard data access interface', 1998.

ISO 10303–23, 'Industrial automation systems and integration – Product data representation and exchange – Part 23: Implementation method: C++ language binding to the standard data access interface', 2000.

ISO 10303–24, 'Industrial automation systems and integration – Product data representation and exchange – Part 24: Implementation method: C language binding to the standard data access interface', 2001.

ISO 10303–26, 'Industrial automation systems and integration – Product data representation and exchange – Part 26: Implementation method: Interface definition language binding to the standard data access interface', 1998.

ISO 13584–1, 'Industrial automation systems and integration – Parts Library: Overview and fundamental principles', 1999.

ISO 13584–42: 'Industrial automation systems and integration, Parts Library: Methodology for Structuring Parts Families', 1998.

ISO CD 18629–13, 'Industrial automation systems and integration – Process specification language – Part 13: Duration and ordering theories', 2004.

ISO CD 18629–14, 'Industrial automation systems and integration – Process specification language – Part 14: Resource theories', 2004.

ISO CD 18629–43, 'Industrial automation systems and integration – Process specification language – Definitional extensions: Part 43: Activity ordering and duration extensions', 2004.

ISO DIS 18629–41, 'Industrial automation systems and integration – Process specification language – Definitional extensions: Part 41: Activity extensions', 2004.

ISO DIS 18629–42, 'Industrial automation systems and integration – Process specification language – Definitional extensions: Part 42: Temporal and state extensions', 2004.

ISO IS 15531–1, 'Industrial automation systems and integration – Industrial manufacturing management data – Part 1: general overview', 2002.

ISO IS 18629–1, 'Industrial automation systems and integration – Process specification language – Part 1: Overview and basic principles', 2004.

ISO IS 18629–11, 'Industrial automation systems and integration – Process specification language – Part 11: PSL-Core', 2004.

ISO IS 18629–12, 'Industrial automation systems and integration – Process specification language – Part 12: Outer Core', 2004.

ISO CD 18629–44, 'Industrial automation systems and integration – Process specification language – Definitional extensions: Part 44: Resource extensions', 2004.

Kemmerer S. J., The grand experience, NIST SP 939, 1999.

Knowledge Interchange Format, Part 1: KIF-Core, ISO/JTC1/SC32/WG2, WD, 1999.

Law K. H., 'Process specification and simulation', PSL quaterly progress report, Stanford University, CIFE, 2001.

Miller P. 'Interoperability: what is it and why should I want it?' 2000, available at http://www.ariadne.ac.uk/issue24/interoperability, accessed July 2006.

NIST, 'Interoperability Cost Analysis of the U.S. Automotive Supply Chain (Planning Report #99–1)', 1999, available at http://www.nist.gov/director/prog-ofc/report99–1.pdf

NIST, 'Economic impact assessment of the international standard for the exchange of product model data (STEP) in transportation equipment industries', Planning report #02–5, 2002.

Pierra G., Sardet E., Potier J. C., Battier G., Derouet J. C., Willmann N. and Mahir A., 'Exchange of component data: the PLIB (ISO 13584) model, standard and tools', Proceedings of the CALS EUROPE'98 Conference, 16–18 September 1998, Paris, France.

Pouchard L., Ivezic N. and Schlenoff C., 'Ontology engineering for distributed collaboration in manufacturing', AIS, 2000.

Pouchard L. and Rana O., 'The Role of Ontologies in Agent-oriented Systems', 6th Joint Conference on Information Sciences (JCIS), Computational Semiotics Workshop, Research Triangle Park, NC, March 2002.

Ray S. R. and Jones A. T., 'Manufacturing interoperability', Concurrent Engineering: Enhanced Interoperable System: Proceedings of the 10th ISPE International Conference, pp. 535–540, 2003.

Tesfagaber G., 'Application of PSL to Construction Process Information Specification and Exchange', PhD Thesis, Loughborough University, 2004.

Tesfagaber G., Cutting-Decelle A. F., Anumba C. J., Baldwin A. N. and Bouchlaghem N. M., 'Semantic process modelling for applications integration in AEC', International workshop on information technology in civil Engineering, ASCE, 2002.

Whatis.com, available at http://whatis.techtarget.com, accessed July 2006.

Chapter 8

Integrated product and process modelling for Concurrent Engineering

*Chimay J. Anumba, N. M. Bouchlaghem,
Andrew N. Baldwin and Anne-Francoise
Cutting-Decelle*

8.1 Introduction

This chapter analyses one important stage of the implementation of Concurrent Engineering (CE) in construction, with the elaboration of a common product and process representation of the design and construction information. This development of a common product and process model is done through the description of a project on which the authors have worked. This integrated model proposed in the project defines some fundamental bases on which CE concepts can be developed, since it provides the common elements of the information exchanges among the actors of the construction process.

One of the aims of the work done within the ProMICE project was to elaborate an integrated product and process model for life cycle design and construction of steelwork structures, enabling the introduction of CE concepts.

In the first part of this chapter, we present an important stage towards the implementation of CE concepts in construction, the identification of product and process related information.

This section is followed by a second part, describing the ProMICE project and the integrated product and process model resulting from the work done within the framework of the project. Then we analyse the way of introducing CE concepts into the ProMICE model. This chapter ends with the presentation of some results obtained at the end of the project.

8.2 Product and process modelling

8.2.1 Types of information models used in the building and construction sector

We focus here on the two main categories of models used in the building and construction sector, product models and process models, since they probably represent the most important way of structuring of the information circulated, handled, exchanged and archived all over the building life cycle.

8.2.1.1   Product modelling

Some years ago (Anumba et al., 1998), most research focused on CAD integration (Anumba and Watson, 1991), considered as the first and most crucial stage towards computerisation in construction. Prototypes were built from different computer techniques, including expert systems, CAD and technical tools (Fenves et al., 1990) at Carnegie-Mellon University (Fisher et al., 1991) at Stanford University, etc. In Europe, several research calls were opened to IT in Construction, and at this time STEP (ISO 10303 STEP standard) (STEP-1–94) was emerging as a major potential contribution to industrial information exchanges. Most projects of that period, such as COMBINE (Dubois et al., 1995), COMBI, ATLAS (Atlas, 1992), CIMSTEEL, etc. built upon this approach and claimed to be 'STEP-compliant'. Many models have been developed for specific or integration purposes; some of them overlap more or less completely, while others are incompatible.

Tools and techniques used for product modelling   Several methodologies (or languages) can be used for product data modelling, among which are: IDEF1×, NIAM, UML, EXPRESS, EXPRESS-G, etc. Some of their main features are presented below:

*   IDEF1×: poor representation of relationships between entities, no temporal representation;
*   EXPRESS: object oriented, no process/dynamic representation, new version imminent;
*   NIAM: not suitable for process modelling, easily understandable, not able to represent dynamics of the model, lack of software tools;
*   UML: the most recent, suited to product and process modelling, object oriented.

Some elements of a (rough) comparison between these methods are provided in Table 8.1, which shows a cross representation of their features and/or properties.

New trends   For years, researchers (Cutting-Decelle et al., 1997) have been pushing the idea of integrated data and process models as a basic step to boost the computerisation of the Building Industry. Since then, research is developing towards the applications needed by the industry, such as CE and electronic document management.

Another formalism, Common Object Request Broker Architecture (CORBA) is proposing a standardised approach of the object-oriented side

Table 8.1 Matrix of features of product modelling methods (ProMICE)

Criteria	IDEF1×	NIAM	EXPRESS	UML
Modelling approach	Relational	Relational and object extension (specialisation)	Object-oriented (OO)	Object-oriented (OO)
Software availability	Poor	No	Yes	Yes
Standardisation	Yes	Yes	Yes	Yes
Ease of use	Fair	Good	Good	Yes/no
Understandability		Good		Yes/no
Dynamics aspects	No	No	V1: no V2: yes	Yes
Expression of constraints	No	Yes	Poor	Yes
Process representation	No	No	V1: no V2: yes	Yes
Entity behaviour	Poor	No	Rules	Yes
Applicability		Analysis for data management		OO analysis and programming

enabling portability and interoperability of heterogeneous systems over heterogeneous networks. The importance of CORBA should not be underestimated because this is a de facto international standard that provides functionalities highly required for the implementation of information exchanges within virtual enterprises – the reality of most of the construction teams. Today, important research axes are developing around data modelling, notably dynamics modelling, semantics and knowledge modelling.

An important initiative is provided by the work done within the (IAI) International Alliance for Interoperability, whose aim is to 'Define a Universal Language for Collaborative Work in the Building Industry.' IAI intends to define Industry Foundation Classes, from which any building at any stage can be described in a common way – paving the way for integrated CAD data environments (IAI, 1997). Launched by Autodesk, it now includes a large number of software developers and industrial interests.

8.2.1.2   Process modelling

The concept of process lacks a commonly agreed definition. A typical definition is 'a set of partially ordered steps intended to reach a goal' (Humphrey, 1992 as quoted in: Koskela, 1995).

There are four common perspectives to processes (Curtis, 1992):

• Functional: representing what process elements are being performed, and what flows connect these elements
• Behavioural: representing when process elements are performed, and how they are performed through feedback loops, iteration, decision making conditions, etc.

- Organisational: where and by whom process elements are performed
- Informational: a perspective of the informational entities produced or manipulated by the process.

In the functional view, processes consist of activities, that together achieve the purported goal. In addition, auxiliary concepts such as artifacts (products of activities) can be used for process representation. In a behavioural perspective, processes may consist of precedence relations or information and material/information flows, with the time explicitly represented. Flow process concepts focus on what happens to material and information in timeline. In an organisational perspective, processes may consist of agents (performing activities) and roles (set of activities assigned to an agent). Also, the process may be viewed as composed of a supplier-customer partnership. In an informational perspective, processes consist of data, objects, documents, etc.

In principle, these perspectives, when combined, produce a complete model of a process. However, in current practice of process modelling, the functional perspective (as provided by SADT method) often dominates: activity is seen as the basic construct, and this process concept only achieves one goal, 'how to obtain the result?'.

Of course the answer to this question is sufficient for achieving the process; however, it does not exhaust all improvement potential. There are two other relevant goals, that should generally be tackled: how not to consume unnecessary resources (Koskela, 1995) and how to ensure that the result corresponds to requirements. In order to achieve these goals, contributions from behavioural and organisational perspectives are needed.

An approach of the construction process According to (Björk, 1992), information handled during the construction process can be divided into several categories:

First, information must state facts: such as design documents, which are the results of design decisions. Information to be transferred between computing systems in the construction process is mostly of this type. This information has also to define goals and requirements which a particular project must fulfil. The third category of information states rules which restrict facts, but which apply in general and are not tied to a particular project. These three categories of information can be called *facts*, *constraints* and *knowledge*. From a programming language point of view, facts can be constructed using assignment statements, requirements are mainly represented by inequality operators (or algorithms) and knowledge through knowledge based systems;

The second point provides a semantic approach dividing information into project-specific and more general information. Facts can be both project specific and general. Constraints are mainly project-specific and knowledge is usually general in nature;

The third point of view concerns the presentation and categorises the types of documents used to present the information for human interpretation. Some typical presentation formats used in construction are: drawings, schemas, realistic visualisations, written specifications, calculation results, bills of materials, contracts, orders and various tendering documents.

The study is here limited to project-specific information, focusing on the semantics of the information. The reason of this choice comes from our primary concern to study information management within construction projects. The information to be communicated to other parties in the construction process mostly consists of factual information. Clearly constraints are very important in the early briefing stages of projects and in quality assurance applications. Knowledge mainly resides in application programs and its effect on the actual transfer of data between project participants will need to be examined further.

Modelling of the construction process Several process models have been developed in the domain of construction, among which the MoPo model (Cooper, 1998), covering the whole construction life cycle; other models mainly focus on the design stage such as the ADePT model described in (Austin, 1996). Some process models introduce CE features, such as the model presented in (Anumba, 1996), or client requirements (Kamara, 2000).

Researchers have developed a number of process models for different stages of the construction process. For example, Austin *et al.* (1996) present a data flow model for planning and managing the building design process; Cooper *et al.* (1998) report on the development of a generic design and construction process protocol; Hannus *et al.* (1997) have developed a prototype tool for construction process modelling and management; and Kamara *et al.* (1998) describe a model for the processing of client requirements in construction.

Tools and techniques Several tools and techniques are available for process modelling. These vary in complexity and functionality often utilising very different formalisms, notations and graphical representations. The ease with which the various modelling constructs can be understood by end-users varies significantly from one approach to another and is an important consideration in choosing between the existing techniques. However, the overriding factor remains the primary objective of the modelling initiative. In many cases, this sets the requirement for the usability of the resulting models.

Existing process modelling methods include the following, only some of which are used in the construction industry (see Table 8.2 for comparison): (ProMICE analysis)

IDEF0/SADT (Structured Analysis Design Technique) – These are almost identical activity modelling tools based on the Input, Control, Output and Mechanism model for representing flows (Hannus, 1992);

Data Flow Diagrams (DFD) – These consist of four basic elements – a data (information) flow, a process, a data store, and an external source (or sink) (Austin *et al.*, 1996);

Role Activity Diagrams (RAD) – Originally developed for software process modelling, this diagrammatic approach focuses on role modelling (Abeysinghe and Phalp, 1997);

Unified Modelling Language (UML) – Based on an integration of the three most prominent object-oriented modelling languages (Booch, OMT and OOSE), UML provides visual models of both products and processes (for more details, see http://www.rational.com).

It should be noted that only the most commonly used process modelling methods have been included in the above summary.

Within the context of the ProMICE project, the IDEF3 process modelling method has not been reviewed, since this method appeared, at the time of the project, not to be widely known in construction.

Trends In addition to the above methods and approaches, several researchers have developed modifications, variants and enhancements to established tools, so as to accommodate their desired modelling perspective. For example, Abeysinghe and Phalp (1997) have combined RAD with Communicating Sequential Processes (CSP) – a formal process modelling paradigm based on concurrency and communication) while Kartam *et al.* (1997) present a 'work-mapping' model that has roots in the conventional system

Table 8.2 Comparison of process modelling methods (ProMICE)

Criteria	IDEF	DFD	RAD	UML
Modelling approach	Static activities	Data flow diagrams	Emphasis on roles; role = sequence of actions and interactions	Object-oriented
Software availability	Yes	Yes	Yes	Yes
Standardisation				Yes
Ease of use	Yes	Yes	Yes	Yes/no
Understandability	Yes	Yes	Fair	Yes/no
Dynamics aspects	No	No	Yes	Yes
Flexibility	Fair	Fair	Fair	Yes
Link to data model	Yes	Yes	No	Yes
Layering	Yes	Yes	No	No
Applicability	Functional modelling	Data flows	Software process modelling	OO analysis and programming
Actor prerogatives	Limited (mechanisms)	No	Yes	Yes

conversion model but incorporates some features of SADT. It is recognised that these variants and hybrid models extend the capabilities of existing methods. However, process models still represent only a partial representation of the development of an artefact as they usually hold no information on the end-products of sub-processes or the final end-product of the overall process.

8.2.2   Synthesis of the approaches towards an integrated product and process model

It is interesting to make a synthesis of the common features of these models, leading to a generic process representation, thus contributing to the overall development of an integrated product and process model for addressing the on-site management phase of a construction project (Kimmance, 2000). The basic role for models of product and process information is to introduce the major elements found within the construction domain. These elements may include the physical or logical entities (object classes) that make up the final products, in essence the overall facility itself, its system and components; the resources used such as equipment and materials; actors and organisations; and information relating to contracts, controls, schedules, etc. The basic entity for representing on-site construction processes is a process or activity entity.

Almost all of the models reviewed in our analysis contained (in some form or another) one or more of these physical, logical and process entities, though they are not often arranged in a toplevel hierarchy. Outside these main entities, various models included entities such as cost, time or quality that, while less universally adopted, may equally belong at the top level. Although the major elements are quite consistent across some of the models reviewed, the basic relationships among these elements varied. Of particular interest is the way in which products and processes are related. One approach was to adopt a simple relationship between products and processes by integrating them, so that processes correspond to certain products. Another approach is to draw upon the basic perspective that processes have inputs and outputs, as embodied in the models utilising IDEF0 techniques, where these inputs and outputs are various product entities.

All of the process models encompass similar features in the form of high-level type processes, divided into sub-level type processes, which can be grouped into categories, such as activities (tasks or events), results or resources. An activity uses resources to produce results. Traditionally construction classification systems often tend to equate results to buildings and their parts, because of the need to distribute total building construction costs over building parts, which is co nvenient for cost analysis purposes. Although, it is evident that information (mostly delivered as documents) and services are other important sub-types of results (Cutting-Decelle *et al.*, 1999).

8.3   The ProMICE integrated product and process model

8.3.1   The ProMICE project

Product and Process Models Integration for CE in Construction (ProMICE) is a collaborative research project between the Department of Civil and Building Engineering at Loughborough University, the UK and the Ecole Supérieure d'Ingénieurs de Chambéry, Université de Savoie, France. It was funded jointly by the British Council and the French Government (Anumba *et al.*, 1999).

8.3.2   Objectives of the project

The aim of the project is to compare and link British and French approaches to product and process modelling with a view to developing a generic integrated model based on CE principles. The specific objectives of the project include:

review and comparison of the use of product and process models in the construction industry in Britain and in France;

development of a generic integrated product and process model for design and construction, based on CE principles. The generic model will embody the best features of French and British practice, and as far as possible will be developed as a conceptual model, independent of implementation constraints;

investigation of the requirements for computer-aided design (CAD) and information technology (IT) systems – including virtual reality (VR) – to support the generic product and process model. These requirements will form the basis for a software architecture for the implementation of the model.

The CE framework within which the integration of the product and process models is being undertaken is innovative and incorporates the best features of CE implementation in the manufacturing industry.

8.3.3   Work programme

To achieve the goals defined for the project, the work has been split into five tasks, which are:

identify available models: for data and processes (the UK and France);
identify available representation methods;

agree on common methods, for data and processes;

elaborate a synthesis of the models to produce the generic integrated product and process model;

identify CAD and IT requirements and formulate a software or logical architecture for the generic model.

8.3.4 Applicability

It is intended that the integrated product and process model will facilitate improvements in the construction process, particularly with respect to: collaborative design, project co-ordination, reduction in project duration, reduction in costs, reduction in claims and disputes, and improvements in product quality. The generic model will be applicable to different European countries, many of which have similarly fragmented construction industries.

8.3.5 Areas of potential concurrency during the lifecycle phases of a construction project

The different life-cycle phases of a construction project can be detailed into eight *tracks* (Prasad, 1998), which are: inception and project definition, outline design, structural engineering and analysis, property specifications, cost management, procurement and supply, fabrication, assembly and erection and finally facility management. The track *facility management* is an ongoing coordination track that runs for the full construction life cycle, also providing normal project management functions, tasks sequencing, cooperation and central support to the other tracks. These eight tracks are not unique to a particular construction facility (such as buildings, bridges, roads, factories, etc.). Individual tasks breakdown, their identifying names and time overlaps may differ from project to project. Figure 8.1 represents possible areas of concurrency during these phases. As we will see it later, the focus of the ProMICE project has been put on the design stages of a construction project.

8.3.6 Modelling approach

Following a preliminary review of modelling languages able to represent both product and process information, the project team decided to use the Unified Modelling Language (UML) (UML, 1997), as it offered the potential for achieving the ProMICE objectives (Anumba *et al.*, 1998). UML is not a modelling method in itself, but a modelling notation, or more, a graphical modelling language used to describe, most of the time, software development processes. Constitutive elements of the language are modelling elements and diagrams: UML defines nine diagrams, four of them bringing a *static view* (Class, Object, Component, Deployment diagrams) and five a *dynamic view* (Use Case, sequence, Collaboration, Statechart, Activity diagrams).

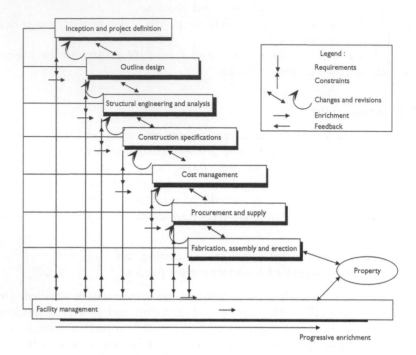

Figure 8.1 Areas of potential concurrency.

It is important to notice that a diagram is not a model, but only a partial graphical representation of some elements of the model: a diagram is a projection onto the model, as a kind of perpective on the model. Several diagrams are necessary to illustrate the entire model.

One of the problems we met when we started the representation of the model with UML was the determination of the types of UML diagrams to be developed and their sequence, since the subject of our development is different enough from the common usage of the language, notably the nature of the system to be described. The system we need to represent (and of which we want to know, the behaviour through the knowledge of elements and diagrams) is made of the design team (architect, engineers, project manager) involved in a building construction project.

For the ProMICE project, we decided to focus our work on the design stage of a construction project, without considering the full life cycle of the building, since this stage can be considered as belonging to the *decisional core* of the construction process. It is a critical stage where inappropriate decisions can have big consequences on subsequent stages, this can be prevented if problems are identified during the early stages of the project.

Compared to software development, the specificity of the use we make of the language lies in the way of defining the specifications of the system: specifications of a building project are known at the beginning, since they have been defined by the project owner.

Activities of the actors involved in the project are defined through activity diagrams and sequence diagrams. These sequence diagrams provide a powerful representation of the sequencing of the different activities, through the description of *working scenario* of the actors involved, thus enabling a detection of possible *strategic crossings* that could be improved using CE features. Figure 8.2 below shows an example of a sequence diagram.

Use case diagrams can be used to provide a high level view on the (main) actors involved in the *system* considered. A rough representation of the outline design stage is shown in Figure 8.3.

The names of all the actions have not been represented on the diagram, for readability reasons. It is interesting to represent at the same time activity

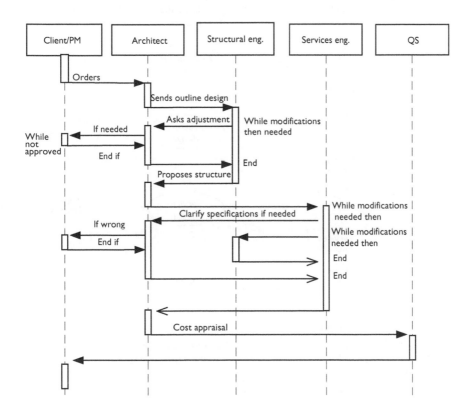

Figure 8.2 Sequence diagram: design stage – traditional approach.

diagrams, since they provide a complementary view, emphasising the flows of control among the actors and their activities. Figure 8.4 shows an example of an activity diagram related to the outline design stage.

For the project, the development of UML diagrams (activity, sequence, use case, collaboration, deployment and state) has mainly been focused on the design stage, involving the actors and the tasks met in a construction project. In order to facilitate the description of the project, not the same according to the nature of the bid or the country, we decided, for a first stage, to separately represent the two models (in France and in the UK). It has then been possible to find a common representation of the project, valid

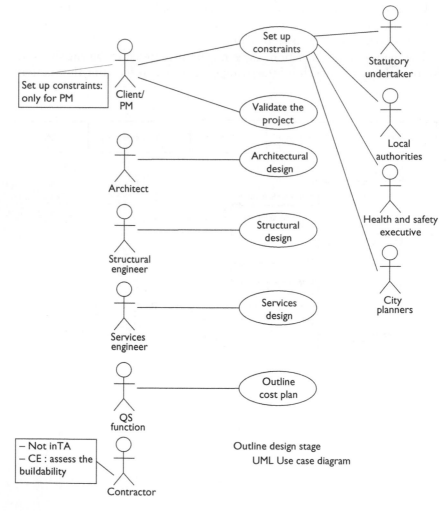

Figure 8.3 Use case diagram: outline design stage.

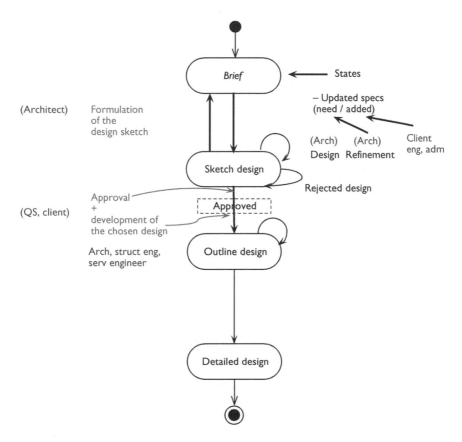

Figure 8.4 Activity diagram: design stage.

for both countries, on which we introduced CE concepts. The validation has been made on a steelwork building project.

8.4   Introduction of CE concepts into the ProMICE model

8.4.1   Methodology

CE features are introduced in the model according to a three-stage methodology we developed for this project.

The aim of this methodology is two-fold: first, we defined the way of working, that is the set of procedures necessary to introduce CE concepts into the model; another feature of this methodology is to provide a way of representing *CE knowledge*, that is how to describe CE specificity in order to introduce the related concepts in the model. In a second stage, it has been necessary to *translate* those concepts into the generic representation provided by the model resulting from the integration.

8.4.2   Stages of the work

The three stages of the method followed in the ProMICE project are:

Stage 1: description of the current situation (traditional approach) in terms of the actors involved in the construction process and in terms of the information flows;

Stage 2: description of a CE way of working (using the same tools as in the stage 1);

Stage 3: defining changes to facilitate the transition from the current situation to a CE way of working.

8.4.2.1   Stage 1: Current situation, traditional approach

This stage used decisional tools, such as behavioural graphs and templates to be completed for each actor at each stage of the design-construction process, nonetheless restricted, for the analysis, to the design stage. The first template was used to define the functions included in the design process at every stage from inception to scheme design (Table 8.3).

The actors' involvement and responsibilities at every stage are then shown on another set of templates using four levels of involvement (None, Low, Medium and High) and three classes of responsibilities (None, Partial and Total), an example of this is shown in Table 8.4. At this stage, it is important to mention that all the diagrams represented already result from a synthesis of the structure of a construction project between the two countries involved in the work.

8.4.2.2   Stage 2: CE approach

The same working procedure is then applied to CE approach of the same construction project. The same decisional tools are used: matrix representation and forms (same as for traditional approach). The result of the matrix analysis is also available on a table showing the actors and the stage of their intervention.

8.4.2.3   Stage 3: Transition from traditional to CE approach

This stage has not been fully developed during the project, for time reasons. The aim of this stage was to make clear the main points targeted by a

Table 8.3 Definition of functions in the design process

Stages function	Inception	Feasibility	Outline proposal	Scheme design
Project initiation	Examine the present circumstances and consider the need to build. Set up project team	Conduct user studies, and provide further information. Consider feasibility report and develop brief	Receive and appraise designs and reports. Approves costs and makes decision to proceed	Approve full design and costs. Authorise formal approval for statutory consent
Management	Liaise with client and obtain background information, budgets, requirements and time tables about the site	Survey and site study and locality. Consult statutory authorities. Prepare feasibility report, site meetings	Co-ordinate the development of the outline proposal and amend brief. Report to client	Co-ordinate design and prepare full scheme and report to client. Apply for planning consents
Architectural design	Discuss terms of appointment: service provided basis of fees	Carry out site studies. Attend meeting, assist in preparation of the report. Obtain outline-planning consent	Carry out outline proposals and contribute to meetings and preparation of report	Prepare full scheme design and pass drawings to QS. Prepare draft report
Structural design	Discuss terms of appointment: service provided basis of fees	Carry out studies on site. Obtain additional information. Contribute to meetings and assist in feasibility study	Contribute to meetings and carry out further studies. Prepare outline design proposals	Assist QS in finalise cost plan, and contribute to scheme design and report
Services design	Discuss terms of appointment: service provided basis of fees	Carry out studies on site. Obtain additional information. Contribute to meetings and assist in feasibility study	Contribute to meetings and carry out further studies. Prepare outline design proposals	Assist QS in finalise cost plan, and contribute to scheme design and report

(Table 8.3 continued)

Table 8.3 Continued

Stages function	Inception	Feasibility	Outline proposal	Scheme design
Costing	Discuss terms of appointment: service provided basis of fees	Obtain additional information. Attend and carry out further meetings and assist with feasibility studies, building cost and tenders	Contribute to meetings studies. Prepare outline cost proposals and plan	Develop and finalise cost plan. Contribute to report
Production	Discuss site operations and running of site	Assist in preparation of feasibility report, and attend meetings and liaise with client	Contribute to the preparation of the report and advise on buildability	Assist in building schedules and advise on buildability
Operation	Discuss terms of appointment: service provided basis of fees	Carry out studies on site. Obtain additional information. Contribute to meetings and assist in feasibility study	Contribute to meetings and carry out further studies. Prepare outline design proposals	Liaise with client, QS and engineers to help with the preparation of the final report
Decommissioning and demolition	Consider life cycle and duration of building and occupants	Obtain additional information. Contribute to meetings and assist in feasibility study	Obtain further information. Contribute to meetings and assist in feasibility study	Liaise with client, QS and engineers to help with the preparation of the final report

Table 8.4 Actors' involvement and responsibility, feasibility stage

	Actor	Client	Project manager	Architect	Structural engineer	Services engineer	Quantity serveyor	Contractor	Facilities manager
Project	inv	High	Low	None	None	None	None	None	None
	resp	Total	Partial	None	None	None	None	None	None
Management	inv	Medium	High	None	None	None	None	None	None
	resp	Partial	Total	None	None	None	None	None	None
Architectural design	inv	Low	Low	Medium	None	None	None	None	None
	resp	Partial	Partial	Partial	None	None	None	None	None
Structural	inv	Low	Low	None	Medium	None	None	None	None
	resp	Partial	Partial	None	Partial	None	None	None	None
Services	inv	Low	Low	None	None	Medium	None	None	None
	resp	Partial	Partial	None	None	Partial	None	None	None
Costing	inv	Low	Low	None	None	None	Medium	None	None
	resp	Partial	Partial	None	None	None	Partial	None	None
Production	inv	Low	Low	None	None	None	None	Medium	None
	resp	Partial	Partial	None	None	None	None	Partial	None
Operation	inv	Low	Low	None	None	None	None	None	Medium
	resp	Partial	Partial	None	None	None	None	None	Partial
Decommissioning and demolition	inv	None	None	None	None	None	None	None	None
	resp	None	None	None	None	None	None	None	None

transition process from a traditional approach of a construction project towards a CE one.

A comparison between the two sets of diagrams and corresponding glossaries (traditional and CE), added to the actor/stage matrices and the related forms enabled the identification of some crucial points of the design process:

• differences between the ways the actors work;
• gaps or overlaps of the function(s) assumed by the actors;
• leading to misunderstandings or lacks of communication.

Athough not fully developed, the results from this stage has highlighted the different problems associated with the design process.

8.5   Results

Among the different results provided by the ProMICE project, we have to separate the results coming from the first two stages of the methodology, from the results of stage 3. Results from stages 1 and 2 enable a more direct (or immediate) validation on a real test case such as a steelwork building: a comparison between the two ways of working as selected for the project seems at a first glance easier to do. Results from stage 3 need further developments in order to really validate the set of rules developed: in that sense, it may appear as a more long term action.

8.5.1   *Results expected from stages 1 and 2*

The analysis of the results of the first two stages enables a comparison between traditional and CE approaches of the design-construction process, but also a comparison between the UK and French ways of working.

Comparison between traditional and CE approaches: The differences between the two approaches clearly appear on the matrices and the forms, but also on the UML diagrams – even if not fully developed as they are today. The differences seem to lie in the important number of *messages* exchanged among the actors in the traditional structuring of a construction project. Besides, those messages are essentially sequential, thus contributing to increase the problems met when something occurs at the end of the exchange process.

Comparison between the UK and French project procedures: The model built up within the framework of the project resulting from a synthesis of the working procedures of the two countries, problems may appear when the model is applied to a French construction bid. To develop the example, we tried to take the most similar type of construction project (in France and the UK), that is the design and build project. Some other types of projects proved to be more or less incompatible among the two countries.

8.5.2 Results expected from stage 3

At the end of the development process, the third stage can enable the elaboration of a set of rules, both for the actors (defining their role) and for the information flows (defining the type of information management to be dealt with by the actors).

This set of rules can be seen as a '*guideline*', providing the way of moving from a traditional project organisation towards a CE one. Of course, these rules need several (industrial) validations, to refine the values of the different parameters.

8.5.3 Industrial validation of the final model

One of the objectives of the ProMICE project was to identify the changes needed by this transition from traditional towards CE approach, then to represent those changes, notably in the domain of steelwork construction and to write guidelines to help users. The objective is also to provide an industrial validation of the final model. This validation has been made on a steelwork building, chosen since this type of construction provides a better *traceability* of the work done by the different teams involved in the project. It has also enabled us to rely on several results (in terms of communication and information exchanges) coming from the Eureka EU130 CIMSteel Project (Cimsteel) on which one of the ProMICE partners has worked for many years.

8.6 Conclusions

At the heart of any good outline design, construction and procurement process, there lies a set of underlying principles for satisfying the interests of the clients, the contracting body, and the company.

This chapter focused on the presentation of these principles, in a context of CE applied to construction, allowing the construction project teams to formulate significant outline design and construction process strategies. The introduction of these CE concepts has then been presented through the work achieved within the framework of the ProMICE project, both in terms of the methodology carried out in the project and in terms of on-going work. The final stage of the work was to represent and validate the changes needed for the transition from the traditional process to CE using the case of a steel frame building.

The work done within this project has been used as a basis for the LEXIC project, funded by the EPSRC at the University of Loughborough, the UK. This project, launched in 2000 for a period of two years, was aimed at developing a common language for the representation and exchange of process information, since it was (and it is still) considered as a important contribution to the integration of concurrent engineering in construction.

8.7   Acknowledgement

The authors would like to thank Professor J. Dufau and Dr M. Mommessin, both from the University of Savoie/ESIGEC (France) for the work done within the framework of the project presented in this chapter.

8.8   References

Abeysinghe, G. and Phalp, K. (1997), 'Combining Process Modelling Methods', *Information and Software Technology*, Vol. 39, pp. 107–124.

Anumba, C. J. and Evbuomwan, N. F. O. (1996), A Concurrent Engineering Process Model for Computer-integrated Design and Construction, in Information Processing in Civil and Structural Engineering Design.

Anumba, C. J. and Evbuomwan, N. F. O. (1997), 'Concurrent Engineering in Design-Build Projects', *Construction Management and Economics*, Vol. 15, No. 3, pp. 271–281.

Anumba, C. J. and Watson, A. S. (1991), 'An Integrated CAD Data Structure for Structural Engineering', *Computing Systems in Engineering*, Vol. 2, No. 1, pp. 115–123.

Anumba, C. J., Cutting-Decelle, A. F., Baldwin, A. N., Dufau J., Mommessin, M. and Bouchlaghem, N. M. (1998), Integration of Product and Process Models as a Keystone of Concurrent Engineering in Construction: the ProMICE Project, Proceedings of 2nd European Conference on Product and Process Modelling, Amor R. (ed.).

Anumba, C. J., Cutting-Decelle, A. F., Baldwin, A. N., Dufau, J., Mommessin, M. and Bouchlaghem, N. M. (1999), Introduction of Concurrent Engineering Concepts into an Integrated Product and Process Model, Concurrent Engineering in Construction Conference.

ATLAS (1992), Architecture, Methodology and Tools for Computer Integration in Large Scale Engineering, ESPRIT Project 7280, Technical Annex Part 1, General Project Overview.

Austin, A., Baldwin, A. and Newton, A., March (1996) 'A Data Flow Model To Plan And Manage The Building Design Process', *Journal of Engineering Design*, Vol. 7, No. 1, pp. 3–25.

Björk, B.-C. (1992), 'A Unified Approach for Modelling Construction Information', *Building and Environment*, Vol. 27, No. 2, pp. 173–194.

CIMSTEEL, Computer Integrated Manufacturing of Constructional Steelwork, http://www.leeds.ac.uk/ civil/research/cae/cae.htm

Cooper, R., Kaglioglou, M., Aouad, G., Hinks, J., Sexton, M. and Sheath, D. (1998), The Development of a Generic Design and Construction Process, Proceedings European PDT Days, pp. 205–214.

Curtis, B., Krasner, H., Iscoe, N., September (1992), *Process Modelling*, Communications of the ACM, Vol. 35, No. 9.

Cutting-Decelle, A. F., Dubois, A. M. and Fernandez, I., (1997), 'Management and integration of Product Information in Construction, Reality and Future Trends', *International Journal of Construction Information Technology*, Vol. 5, No. 2.

Cutting-Decelle, A. F., Anumba, C. J., Baldwin, A. N., Dufau, J., Mommessin, M. and Bouchlaghem, N. M., Introduction of Concurrent Engineering Concepts into an Integrated Product and Process Model, Proceedings of the 2nd International Conference on Concurrent Engineering in Construction (CEC, 99) Conference, Espoo, Helsinki.

Dubois, A. M., Flynn, J., Verhorf, M. H. G. and Augenbroe, F. (1995), Conceptual Modelling Approaches in the COMBINE Project, Final Combine Workshop Paper, Dublin.

Fenves, S. J., Hendrickson, C. and Maher, M. L. (1990), Integrated Software Environment for Building Design and Construction, Computer Aided Design, Vol. 22.

Fisher, M. A. (1991), Using Construction Knowledge During Preliminary Design of Reinforced Concrete Structures, Phd Thesis Stanford University.

Hannus, M. (1992), Information Models for Performance Driven Computer Integrated Construction, Proceedings CIB W78 Workshop on Computer Integrated Construction, Montreal, 12–14 May, pp. 258–270.

Hannus, M., Lahdenpera, P. and Vahala, P. (1997), Prototype Tool for Construction Process Modelling and Management, Concurrent Engineering in Construction, Anumba, C. J. and Evbuomwan, N. F. O. (eds), Institution of Structural Engineers, London, July, pp. 65–76.

Humphrey, W. S. and Feiler, P. H. (1992), Software Process Development and Enactment: Concepts and Definition, Technical report SEJ-92-TR-4, Software Engineering Institute, Carnegie-Mellon University, Pittsburg, PA.

Industrial Automation Systems and Integration – Product Data Represen-tation and Exchange (1994), ISO IS 10303–1, Part 1: Overview and Fundamental Principles.

International Alliance for Interoperability (IAI) (1997). IFC version 1.5.

Kamara, J. M., Anumba, C. J. and Evbuomwan, N. F. O. (1998), A Process Model for Client Requirements Processing in Construction.

Kamara, J. M., Anumba, C. J. and Evbuomwan, N. F. O. (2000), 'Developments in the Implementation of Concurrent Engineering in Construction', *International Journal of Computer-Integrated Design and Construction*, Vol. 2, No. 1, pp. 68–78.

Kartam, S., Ballard, G. and Ibbs, C. W. (1997), 'Introducing a New Concept and Approach to Modelling Construction', *ASCE Journal of Construction Engineering and Management*, Vol. 123, No. 1, pp. 89–97.

Kimmance, A. G., Anumba, C. J., Baldwin, A. N., Bouchlaghem, N. M. and Cutting-Decelle, A. F. (2000), Integrated Models for Product and Process

Information: A Keystone of Concurrent Engineering for Site Management, CODATA 2000 International Conference, Baveno (I), October 2000.

Koskela, L. (1995), *On foundations of Construction Process Modelling*, CIB 78, Standford Univ, Palo Alto, USA.

Prasad, B. (1998), Towards Applying Principles of Concurrent Engineering for Efficient Design and Development of Construction Facilities, *Submission to CIDAC Journal*.

Unified Modelling Language, V. 1.1, Rational Software, 1997.

Chapter 9

Document management in concurrent life cycle design and construction

Robert Amor and Mike Clift

9.1 Introduction

In current engineering practice, be it concurrent or otherwise, documents are the central mechanism for communicating, informing and instructing. Any attempt to engender a greater uptake of concurrent engineering (CE) in the industry has to recognise the central role of documents in process re-engineering. The proper management of documents has the potential to greatly improve the design process in terms of efficiency and effectiveness. It is estimated (by document management system developers) that professionals in the industry spend 30 per cent, or more, of their time managing documentation in current paper-based management regimes, and the source of much litigation in the industry can be tracked back to improper management of documentation. IT-based approaches can greatly impact on document management; however, to date the various aspects of IT applied to engineering have developed independently, leading to stand-alone product, process and document management systems. This development path, though productive in each individual area, misses the major gains that can be achieved from integration of all aspects of IT usage. This chapter shows that the proper management of documents provides information about all aspects of a project. It is argued that, through careful management, documents can provide the means to effectively co-ordinate work on the activities required to complete a project, and to determine how processes can be managed to greatest effect using CE frameworks.

9.2 What is a document?

The meaning and scope of the word 'document' in the traditional design and construction process was always clear. A document was any collection of paper that related to a project. Several distinct styles of data layout were used to present information in these documents (see Section 7.2). The physical nature of paper helped define its contractual and legal status as a document, but as we move into an age where an electronic representation of all traditional paper documentation is possible, the definition of a document becomes blurred. Some paper documents may not currently be able to be stored electronically (e.g. standards) and some forms of information that are sent electronically were not considered documents in traditional practice (e.g. e-mail or telephoned orders).

To help define the scope of electronic documents we use a modification of the paper document definition to specify that:

> An electronic document records any transfer of information which occurs during a project, providing views of the project's product model and supporting all processes in a project.

This greatly broadens what can be considered as a document. It still covers everything which has a paper form, but also extends it to information transfers such as:

- phone calls
- a colleague's or third party advice
- project discussions
- data files used for a design tool
- video clips of the construction site.

9.2.1   Document types required in construction

There are a number of documents traditionally generated and stored in paper form for construction projects. Their source and regarded status is as shown in Table 9.1.

For construction projects in which the design has often not been completed until hand-over (or beyond), forms of day-to-day communication need capturing. These include those shown in Table 9.2, all of which have contractual implications.

9.2.2   Documents as the industry's knowledge base

Currently, documents and their management form the basis of the construction industry's knowledge base. All information about a project resides in documents. Decisions about a project (e.g. standards constraints, appropriate products) are based on information found within documents. A firm's library and the available published documents form the knowledge base from which the industry operates. The mark of a successful project is often how well this vast store of documents is managed, both within an enterprise and across enterprises on a project.

Table 9.1 Standard documents in a construction project (the UK viewpoint)

Type	Author	Legal/contractual status
Brief	Client/owner	High
Contract/commission	Client	High
Drawings	Designer/contractor	High
Specifications	Designer	High
Bills of quantities	Quantity surveyor	Medium
Tender documents	Designer	High
Valuations	Quantity surveyor	Medium
Payment certificates	Designer	High
Program/schedules	Contractor	High
Calculations	Designer/contractor	Medium
Site diaries	Supervisor/contractor	Medium
Change orders	Client/designer/contractor	High
Progress records	Supervisor/contractor	Low
Claims/compensation events	Contractor	High
Letters	All	High
E-mail	All	Low
Fax	All	Low
Request for information	Contractor	Medium
Confirmation of instruction	Designer	Medium
Notices	Client	High

Table 9.2 Communications in a construction project

Type	Author	Legal/contractual status
Phone	All	Low
Verbal order	Client/designer	Medium
Advice	All	Low
Video/progress photos	All	Low

9.3 Paper-based document management

In current practice, documents are usually distributed, and held, in paper form requiring a costly and resource-intensive filing, retrieval and issue system. It is estimated that 30–40 per cent of an engineer's work effort on a project is concerned with the management of documents. The scope and layout of documents are determined by practice and may be laid down in standards which differ from project to project, company to company and, of course, country to country.

Construction projects are characterised by their one-off nature and the rapid assembly and disassembly of the project team and their members, many of whom join during the project and depart before its completion. Many construction organisations are very small (1–2 persons) and may only be formed for the particular project, these organisations have in the main not made the transition away from paper-based systems. Even larger organisations have failed to embrace the idea of electronic document management and interchange, mainly because there is no industry accepted standard, there is incompatibility between systems adopted by organisations, and penetration is limited so that paper versions are anyway needed at all times.

Current practice therefore is for all documents, even those that have been generated or transmitted electronically, to be stored as paper for distribution and legal or contractual purposes. Individual firms then run different management processes for the documents pertaining to a particular project usually associated with a particular classification system to enable efficient retrieval at a later stage. This process is cumbersome and error-prone, explaining many of the problems in the industry with out-of-date versions, missing documents, etc.

9.4   IT-based document management

From very early in the commercial use of computers document generation and management has been a major function to be supported. The visionaries foresaw a world where physical documents were not required (with advertising slogans around the 'paperless office') and then as work was undertaken on product modelling systems they also prophesised the demise of documents themselves. As is often the case the visionaries' expectations of computers were far in excess of their capabilities and didn't take into account human factors in the application domain of their computer systems. It is clear that physical documents will not disappear and that documents as an information transfer mechanism on projects will be with us for many decades yet.

However, IT-based document management systems (DMS) can support many tasks on a project and there is a range of functions available in current DMS which would provide benefit for the construction industry in the short term. See for example, Wager and Winterkorn (1998) who present a summary of over 45 DMS categorised by potential functionality, along with user surveys and case studies, or Laiserin (2001) for a more recent view of surviving products and strategies in the market. However, as there is a very low uptake of DMS in the industry these benefits are not realised. The short term benefits which could be realised by utilisation of a DMS (a general overview is found in Laqua, 1999) are mostly in the automation of non-value adding processes, for example, automatic forwarding of documents to a set of team members on completion of a particular process or activity. Standard functions could also automate many of the tracking and verification activities required for dispute resolution by recording who received what documents, at what time, and by recording when the recipient opened the document. Standard functions will also allow security to be implemented through digital signatures to ensure that original versions can be identified and encryption should be used to ensure that unauthorised access to documents can be controlled. For example, Mokhtar and Bedard (1994) proposed a central database as the source of all technical documents to reduce the estimated 50 per cent of problems in buildings which arise through decisions and actions taken in developing working drawings. They envisaged such a system helping through the production of integrated documents, through quicker communication of documents and through the production of document types which span disciplines.

Although an electronic document records any transfer of information it need not contain the full content of the information transfer. For example, an electronic document representing a verbal order is likely to contain the essence of the order rather than the whole audio capture of the conversation (though this may be recorded if required). In many cases the electronic document may capture a reference to existing paper documents which were utilised during the project, for example, a firm's collection of printed codes and standards.

Related research projects include Turk (1994), Björk (1994) and Turk *et al.* (1994). These projects operated on the assumption that, as a preliminary means to achieving Computer Integrated Construction (CIC), a construction document management system (CDM) could be constructed. The CDM is seen as a short term solution, being replaced by full product management systems when the technology matures. A strong case is put for the need for explicit research work on development of CDM systems for the integration of documents within single projects across organisational boundaries. However, it is clear that different disciplines' views of a proposed building are irreconcilable with computerised tools, so even an integrated product database will not provide the ability to reconcile all views.

9.4.1 Functionality that can be supported

There are a number of fundamental activities that a DMS is capable of supporting, these include the following:

- *Storage and retrieval of documents.* Providing access to project participants to deposit or retrieve documents into the system. Can be tied with security systems to ensure only appropriate project members manipulate documents in the system.
- *Notification of document updates.* Linking process management into the handling of the documents to ensure that particular events trigger messages to particular members of a project team. This ensures

tracking to identify who was notified of modified documents (e.g. a signed-off document).

- *Compatibility with paper systems.* Allowing a mix of electronic and paper documents to be incorporated in, or referenced from, the same system.
- *Common file formats or error-free conversion methods.* To allow all document types to be viewed or transformed. An industry accepted standard would greatly assist setting protocols for each new project and team.
- *Document access management.* Access is to be recorded and identity noted; whilst modification to a document should not be possible, it may be used as a template for a new version and that action recorded.
- *Version management.* Latest and earlier versions must be retrievable and must be able to determine the history of document revisions and those project participants involved.
- *Search functionality.* Full text search functionality in text based documents must be possible. The descriptive data sets identifying a document (meta-data) should be searchable and include all necessary information such as type, project, contents, originator, date (created/ sent) etc.
- *Electronic signature.* An electronic signature will need to satisfy legal and contractual issues.
- *Markup.* Annotation, such as red-lining, can be applied to documents within the DMS.
- *Shared database.* Other users must share the document management system databases.
- *User friendliness.* A new user to the system should only require a short training period (measured in hours), which should be held on site.
- *File locking.* The system must be able to restrict access to a document when it is under modification, or identify who is currently working on a particular document.
- *Views.* Parts of a document, for example, a drawing, must be capable of being viewed without the need to call up the whole document.
- *Legal aspect.* The document must be retrievable in such a way that the legal requirements for its authenticity are fulfilled, for example, data format and visualisation software have to be formally specified and linked to the document content.

9.4.2   Different models of IT-based document management

DMS have been developed in several major forms. This section examines the major features of the main categories of DMS to detail the benefits they provide in a project and the drawbacks associated with them.

9.4.2.1   *Electronic document management systems and*
product data management systems

A range of bespoke and commercial Electronic Document Management Systems (EDMS) systems have been developed for the industry over the past few decades. These systems mostly aim at the internal document management processes that need to be supported by the organisation and hence try to reduce the overhead of document management by their staff and improve the organisation's management of document processes on a project. While these systems can provide benefit within the organisation for document processes they are seldom as effective in use on a project. This is due to the remainder of the project team not having access to the chosen system, or not being happy to discard their own systems to use another system for a single project.

Manufacturing industries rely to a great extent on long term partnerships through an extended supply chain where it is feasible to install compatible electronic document management and transfer systems, and in fact may be a pre-requisite for joining the partnership. The construction industry is, however, more likely to be comprised of short term relationships and sub-contracts, the duration of which may often be shorter than the time taken to produce the product (the constructed asset). Various contracts are continually formed, and disbanded, during the production process of the building, which makes it difficult to ensure that such an approach will be adopted by later parties.

Tangible benefits to these participating organisations, for example, improved product quality, timeliness and lower whole life costs, will encourage the more extended supply chain sub-contractors to embrace the concept, providing there is an industry standard. In fact, many organisations which provide products and components for the construction industry also supply other manufacturing industries more at home with EDM.

The major services offered by EDMS for engineering and construction are: the ability to manage CAD files; document capture through scanning and conversion; folders or cases for handling complete projects; recording document workflow to allow documents to be routed to users; distributed databases for interoperability; integration of other databases and computer networks; high security control; ability to handle large numbers of documents (in the hundreds of thousands); and the flexibility to manage compound documents of various types.

The types of document that can be recorded in electronic form could be maintained in a plain relational database system, which would be managed by all those people concerned with a particular project. However, if a domain specific system is implemented, with knowledge about documents and their usage, a much higher level of functionality can be supplied to the users. This can be seen in the commercial arena by the plethora of EDMS which are available for managing documents, over a hundred of which are aimed at engineering documents. Knowing the nature of the objects that it is dealing with, an EDMS can automate many document management tasks, such as logging document creation and modification times or automatically generating version numbers for modified documents.

Product Data Management Systems (PDMS) have the properties of an EDMS as well as further product specific functionality, the major components of which are: configuration control for products and assemblies; management of relationships between items (e.g. CAD and word processor files); control of variants versus standard products; change impact assessment and management; and improving the flow of application data.

Though there are over seventy-five readily available commercial systems aimed at engineering domains (IIC and Cimtech, 1997), these have had very limited take-up in the construction industry. Most of the commercial EDMS are aimed at individual firms, and the total management of documents inside the firm, rather than at a project level with the recognition of multiple external partners needing to collaborate. The commercial EDMS vendors appear unaware of previous research into integrated design systems in the construction domain so their products tend not to support connections with product models. Most systems treat documents and their management as a totally disparate field from product modelling, though there are enormous overlaps in the information manipulated in both of these areas.

A recent research project tackling the interoperability issues in this area was the EC funded CONDOR project (Rezgui and Cooper, 1998). This project aimed to specify a unified interface to a range of DMS systems so that the problem of multiple systems in use by the organisations coming together for a project could be overcome. The final system looks like a single DMS, though requests related to documents could be propagated to a wide range of systems sited in different organisations.

9.4.2.2   Internet-based DMS

A major new trend, as with many service-based systems, has been to repackage DMS systems into an Internet form. As DMS provide some form of groupware functionality (computer mediated human to human interaction) they benefit from development within a medium which provides open and affordable access to all potential participants in a project. The majority of Internet-based DMS are based upon the same principles, and proffer the same functionality, as the existing EDMS developments. They do, however, have the following important beneficial attributes:

- *Affordable, pervasive and consistent interface.* The availability of freely available commercial web browsers on almost every type of machine allows a unified service to be provided to almost every potential user of the DMS. The service is guaranteed to reach all users across all platforms in the same manner.

- *Simplified training for the DMS.* By utilising the web browser functionality and standard web protocols there is a large reduction in the training which needs to be expended in getting users up-to-speed with the DMS.
- *Interoperability of DMS system components.* As the DMS is based upon Internet protocols there is greater potential to link with related services and provide users with a greater depth of task support than a stand-alone DMS system can hope to achieve.

A major impact in this area has been the emergence of CAD-linked Internet-based DMS from all of the major CAD vendors. These sites have given industry professionals, utilising the same CAD system on a project, the ability to easily establish project specific document management across the whole project team.

The Internet community have also been interested in documents and have developed mark-up languages for the representation of documents in the Internet medium (e.g. HTML, XML, etc.). Though these standards do not currently have the representational power to rival the standards used in construction (e.g. CAD representations) they can certainly represent the meta-data required for a DMS to perform its functions. For example, Zarli and Rezgui (2000) surveyed technologies for documents within virtual environments. They describe a general architecture for the construction of open and dynamic virtual environments, and recommend XML technologies for future developments in this area. DocLink (2002) defines a set of XML-encoded transactions enabling a standardised interface to DMS systems, similar to the CONDOR concept previously described. The XML approaches are also mooted as potential paths to reduce the amount of effort required (and hence pathway for errors) to enter attributes and classifications for all documents within a DMS system.

9.4.3 Legal aspects

The production and storage of documents on computer systems has become common practice in the manufacturing industry (and is becoming so in the construction industry) and will increasingly be used for business transactions such as ordering materials and plant. Codes of practice are a standardised method to ensure consistent and competent application of good practice for a particular purpose. For example, the British Standards Institution (BSI) Code of Practice 'Legal Admissibility of Information Stored on Electronic Document Management Systems' (BS PD0008 1996) covers issues such as systems planning, implementation, initial loading and procedures for the use of the system. It pays particular attention to setting up authorised procedures and subsequently the ability to demonstrate, in a Court of Law if needed, that the procedures have been followed. However, legal or contractual admissibility of electronic documentation cannot be realised without the consent of the contracting partners. They need to feel secure when acting upon electronically transmitted data as opposed to waiting for the paper version to be delivered some time later.

The BSI Code of Practice also notes that image-processed documents are currently treated in the same way as photocopies or microfilm (i.e. as secondary evidence). In the adversarial legal system, the other party may try to discredit the integrity of the electronic document and the system on which it was recorded, as well as to dispute its content. There is a long tradition of trust in paper based documentation and limited knowledge, and therefore limited confidence, in electronic methods. In cases where an electronic document has been submitted, the Court will want to question its history in order to evaluate its validity and evidential weight. This forces organisations to ensure appropriate processes and tracking exist to ensure that their document management system will be recognised as providing the same level of reliability as paper-based processes.

Contracts entered into between the players early on in the process can be readily formed on the understanding that data will be exchanged electronically, perhaps as dictated by the client. Under traditional forms of construction contract, the design and construction phases are carried out by different organisations, each contracted to the same client. In such situations the client can impose a requirement that all data will be exchanged electronically. However, the client normally has little jurisdiction over the main contractor's sub-contractors and it is hard to envisage how the idea of EDM can be taken beyond that being exchanged between the client, the consultant organisations and the main contractor.

Yogeswaran and Kumaraswamy (1997) surveyed the major causes of construction litigation, many of which revolve around unclear and inadequate documentation. They detail how IT can be used to re-engineer contract documentation to help reduce such claims.

9.4.4   Future IT directions

It is likely that IT-based DMS will follow a fairly conservative development path, with more features supported in existing systems, and greater interoperability offered across systems and to related services (Björk, 2003 addresses 10 business research issues). However, a few IT advances provide future technical paths for DMS.

Looking to Internet technologies, the success of systems such as Napster and Gnutella point to the possibility of point-to-point topologies for DMS. This could provide greater control over document management for individual organisations, but also enable them to publish required documents for project collaboration. It also moves away from reliance upon a single centralised document server of one type to a distributed approach which would better suit the plethora of systems hosted by different organisations.

With the increased storage capacity available on PCs and servers within organisations and with compressed sound formats it is easily possible to record all utterances on a project as part of the store of documentation for a project (less than a terabyte is required to retain every conversation a person hears in their lifetime, Bell and Gray, 2001). Though social considerations are likely to influence whether this becomes normal practice, it provides a further aspect which can be incorporated within a DMS.

9.5   Forthcoming roles for document management

As documents are not going to disappear in the foreseeable future it is worth looking to the processes in construction which could be impacted by continued use and integration of the evolving DMS available today.

9.5.1   Combining product, process and document views

Boundaries currently exist between project views of product, process and documents. However, there are inherent relationships between documents, data and process and these are likely to become more intertwined. The first part of the document, process and product triangle comprises the connections between a document and the activities which went into the document creation. Documents are part of the input to the majority of activities and are part of the output of many activities or processes in a project. This idea ties to previous work in the development of generic process maps for construction (BAA, 1996) which are then specialised down to the actual project level with known teams and responsibilities. Current work on data standards provides the representational capability to manage the connections between process and related documents and products.

Figure 9.1 can be used to demonstrate how such a process map can be used either for current working, for checking on past work, or for a look-ahead to future work requirements. Figure 9.1 shows part of the detailed process for a construction project with processes to be completed and the flow of control between the processes. In this view it also shows the documents which feed into each process and those which are created during the process. In this view the user can inspect all of the documents which fed into the process and which came out, for example, a series of Word documents, CAD detail drawings, and a VR mock-up of views when passing through the building.

With an active process map the user is informed of the running processes and can see the status of documents and models (completed, being worked upon, and not started) which are being created or modified during the process. In the future view the user can identify all documents which must feed into a process and determine from which processes these documents

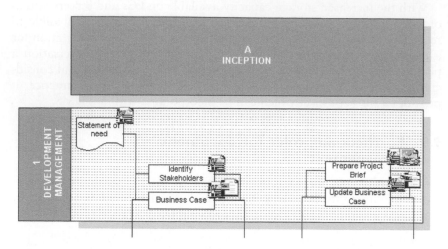

Figure 9.1 A process map specialised for a project with relevant documents.
Source: BAA (1996).

will be created, and hence what still needs to be completed before the examined process could be activated.

These process maps allow top level general information to be viewed and also provide the ability to drill down to very detailed process specifications. At the top levels the system shows the documents which feed into any of the processes encapsulated by a top level general process, and any outputs from the lower level processes which feed into other processes further down the line. That is, these process maps show the document interfaces between processes, either aggregated at higher levels, or in great detail at the more detailed process level.

The final part of the product, process, and document triangle is the possible linkages from product models, and the tools which manipulate product models as part of the design and construction process, to the related documents and processes. To this extent all product information should allow a user to navigate through to the related documents and processes. Current work on standards, especially the IAI-IFCs version 2× (IAI, 2001), provide the representational capability to manage the connections between products and related documents and processes.

For example, in a CAD system it should be possible to select a single element (or a whole sub-assembly) and determine which documents refer to, or impact on, the selected element. These documents would provide information on constraints on the element, for example, signed off specifications, as well as preferences for its design. At a later stage in the building's life, for example, during facility management and maintenance, the documents would provide information on the original specification, tenders, and as-built drawings. This access to the document trail helps to identify who has been (or is currently) working on the element selected and what decisions have been made which may impact on the work that is currently being performed.

In terms of enhanced co-operation and collaboration these linkages provide users at all stages of a project with directed access to all information which relates to various portions of the building, no matter where this information might be stored. This can help ensure that all participants in a project are aware of the decisions which affect their work and are aware of the current set of constraints which limit what they are able to do in their work. In terms of managing liability in a project it ensures that all participants are aware of the constraints on their work and at what stage of finality the portion of the project they are working upon is currently at.

9.5.2   Dispute reduction

Document management in Concurrent life cycle design and construction (CLDC) has the potential to reduce disputes on projects through guaranteed supply of up-to-date information to all project participants. It also introduces new methods to ameliorate legal admissibility considerations for previous problem areas. The major ones are considered below.

Verbal instructions issued on site have always been a breeding ground for legal and contractual wrangles. Contracts usually have a clause providing for a maximum time by when a verbal instruction must be confirmed in writing, often by the receiver of the instruction back to the instructor, who of course can query it. As often as not the work will have been started before the written confirmation has been generated. This gives the person carrying out the work the opportunity of matching or revising the original instruction to the actual work carried out. This, not surprisingly, leads to dispute.

There is emerging technology that should now resolve this in the form of hand- and palm-held computers (or even wearable computers) that can communicate not only with each other via infrared but with head and site office PCs via modem. This means that the whole team can be immediately appraised of the instruction and provide the necessary support, or countermand it if there are wider implications not appreciated on site.

Telephone conversations pose similar problems and require confirmation if a contractual event, such as a request or instruction is included. Actual recordings are not considered viable, they are time consuming to review and are treated with suspicion. However, the fact that a phone-to-phone connection was made, when and for how long is commonplace for billing purposes. Obviously the nature of the conversation is not recorded or even if anything pertinent was said at all.

E-mail and messaging has the advantage of being relatively easy to operate and the system can record that it has been received. It should not be difficult to note that it has at least been read by requiring an acknowledgement.

Site diaries are kept by the client's site representative and are often submitted as supporting evidence in case of disputes. These have traditionally been in hard copy format (like a ship's log) to be inspected by the client's contract supervisor during site visits. Items recorded will include delivery of materials, labour on site, inspections, visitors, site activity, weather conditions, stoppages, etc. Diaries are often backed up with site progress photographs normally taken from fixed locations and on a regular basis, for example weekly or monthly unless specifically requested. A digital camera or video used by an inspector (e.g. mounted in a hardhat) could provide a useful and accessible view for the off-site team and provide supporting evidence for contractual debates.

9.5.3   Supporting health and safety and building handover

In many countries, regulations (e.g. the Construction (Design and Management) Regulations (CDMR, 1994) in the UK) have created new legal responsibilities for clients and their consultants and contractors when undertaking most forms of construction. The construction industry has an unenviable track record when it comes to health and safety and regulations like CDMR aim to improve this, not only during construction, but also when carrying out the maintenance, alteration, refurbishment and ultimate demolition of the building. A Health and Safety file is prepared during the design and construction of the building containing information for safe occupation and is handed over with the building to the owner. This becomes a legal requirement and is accompanied by a considerable amount of drawn and manufacturers' information which is rarely structured in a form that is useful for the facilities manager. Providing mechanisms to allow the appropriate extraction of views of the DMS repository suitable for this purpose then becomes an interesting issue. Clayton *et al.* (1999) examine how to provide facility information automatically structured into documents appropriate to support facility management.

Regular daily photographs linked to activities will also provide a useful record of how the building was put together, where services are buried and how to access or dismantle parts of the building for future maintenance, replacement and refurbishment.

CAD drawings often lose attribute data if they are transferred to other applications such as those operated by the facilities manager. The facilities management application will only use the final versions of design drawings and rarely needs access to earlier versions. Layering the information for facilities management use will ensure the right level of data is transferred at hand-over and hopefully cut out the huge mass of superfluous information.

9.6   ToCEE: An example of advanced document management

As an example of the potential of document management as part of a CE environment we discuss the European Union funded project Towards a Concurrent Engineering Environment (ToCEE) which was completed in 1998 (ESPRIT, 1995; Amor *et al.*, 1997). The primary objective of the ToCEE project was the development of an overall conceptual framework, along with specific software tools, for CE support.

Key issues for a successful CE approach that were addressed are:

* distributed process, product, document and regulation requirements modelling with special focus on intra- and inter-model operability
* inter-discipline conflict management
* legal aspects related to the product data and the electronic documentation
* information logistics and communication management
* monitoring and forecasting
* cost control.

Models were developed by the project under separate work packages to cover process, product, logistics and documents, with cross-cutting themes of legal issues, conflict (clash) management and standards and regulations.

One of the major measures of the benefit of a CE environment is its level of support for co-operation and collaboration between all participants in the project. The ToCEE project delivered an infrastructure which aimed to engender better co-operation and collaboration with a project through the provision of open interfaces to all participants' organisations and also through an open interface to all services provided to the project. This allowed controlled access by any participant to all levels of information in the system, whether that be the evolving product model, the state of the process management system, or the documents held in the project. Figure 9.2 shows the general framework of the ToCEE system in relationship to the clients and services involved. The server end of ToCEE ties together the full set of product, process, document, regulations and conflict management systems from the involved participants to be visible through a single interface. Then from the client's end it is possible to access this unified set of services without regard to the location of any of the services or their information. Figure 9.2 shows the interfaces currently supported in ToCEE, which are via e-mail requests, through an Internet browser, or through tailored wrappers around existing design tools.

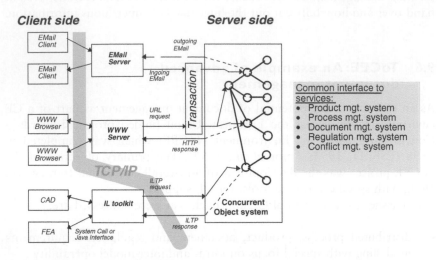

Figure 9.2 The ToCEE system framework.

One of the servers implemented in ToCEE was a DMS managing all project documentation and communications. In this DMS particular emphasis was paid to how such a system can keep track of document version numbers, the interrelationships between documents, and how they could be audited (initial data models and functionality specifications of the DMS can be found in Amor and Clift, 1996 and 1997). This was managed whilst retaining the document character of electronically stored data. The model had to have an open architecture which could be easily adapted by users to meet requirements particular to them. It had to address current and emerging standards for project information, making use of existing systems where appropriate and seeking to influence their future development so that they facilitate the future requirements of electronic concurrent (collaborative) engineering. Although the document model developed was specified independently from the product and process models there were close links between the models to help maintain the legal and auditing requirements of the emerging CE environment.

9.7 Conclusions

It is clear that documents and their management remains a vital role in a successful project. The move towards IT-based document management, and especially Internet-based systems, is proving effective in supporting CE principals, especially enhanced collaboration and cooperation across organisational boundaries for team members in a project. Though the use of

DMS creates some legal issues in terms of processes and admissibility in a legal system based around paper processes there are best practice guides which ameliorate this issue. It is also clear that best-practice usage, and future developments, of DMS will provide for greater life-cycle support especially for facility management. The impact of interoperable systems, based around standard data models, is going to require integration across the product, process and document views which are currently treated separately in system developments in this industry. This remains one of the biggest research issues in effective management of information on a project.

9.8 References

Amor, R. and Clift, M. (1996) Document Models and Concurrent Engineering, in Turk, Ž. (ed.) Construction on the Information Highway, Proceedings of CIB-W78 Workshop, CIB 198, Bled, Slovenia, 10–12 June, pp. 33–34.

Amor, R. and Clift, M. (1997) Documents as an Enabling Mechanism for Concurrent Engineering in Construction, in Anumba, C. J. and Evbuomwan, N. F. O. (eds) Concurrent Engineering in Construction, Proceedings of 1st International Conference, London, UK, 3–4 July, pp. 151–162.

Amor, R. W., Clift, M., Scherer, R., Katranuschkov, P., Turk, Ž. and Hannus, M. (1997) A Framework for Concurrent Engineering – ToCEE, European Conference on Product Data Technology, PDT Days 1997, CICA, Sophia Antipolis, France, 15–16 April, pp. 15–22.

BAA (1996) The Project Process, British Airports Authority, Gatwick, UK.

Bell, G. and Gray, J. (2001) Digital Immortality, Communications of the ACM, 44(3), March, pp. 29–30.

Björk, B.-C. (1994) Conceptual Models of Product, Project and Document Data; Essential Ingredients of CIC, ASCE First Congress on Computing in Civil Engineering, Washington DC, USA, 20–22 June.

Björk, B.-C. (2003) Electronic Document Management in Construction – Research Issues and Results, ITcon, http://www.itcon.org/, Vol. 8, pp. 105–117.

BS PD0008 (1996) Code of Practice for Legal Admissibility of Information Stored on Electronic Document Management Systems, BSi, London, UK, ISBN 0–580–25705–3, pp. 64.

CDMR (1994) Construction (Design and Management) Regulations, HMSO, London, UK.

Clayton, M., Johnson, R. and Song, Y. (1999) Downstream of design: Web-based facility operations documents, Computers in Building: Proceedings of the CAADfutures'99 Conference, Augenbroe, G. and Eastman, C. (eds) Atlanta, USA, 7–8 June, pp. 365–380.

DocLink (2002) Leeds University, http://www.doclink.info/ (accessed July 2006).

ESPRIT (1995) ESPRIT IV-20587 ToCEE: EU ESPRIT IV Project 20587, ToCEE – Project Programme, EU/CEC, Directorate Generale III, Brussels, Belgium.

IAI (2001) Industry Foundation Classes version 2×, http://www.interoperability.com/ (accessed July 2006).

IIC and Cimtech (1997) Engineering Document Management Systems and Product Data Management Systems, 3rd ed., ISBN 0–900458–71–2, 192 p.

Laiserin, J. (2001) AEC Project Webs Redux, CADscope, Vol. 1, 8–14.

Laqua, R. (1999) What is Happening to EDM? Gateway Consulting Group, http://www.gatewaygrp.com/ (accessed July 2006).

Mokhtar, A. and Bedard, C. (1994) Towards Integrated Construction Technical Documents – A New Approach through Product Modelling, Proceedings of the First European Conference on Product and Process Modelling in the Building Industry, Dresden, Germany, 5–7 October, pp. 3–10.

Rezgui, Y. and Cooper, G. (1998) A Proposed Open Infrastructure for Construction Project Document Sharing, ITcon, http://www.itcon.org/, Vol. 3, pp. 11–25 (accessed July 2006).

Turk, Ž. (1994) Construction Design Document Management Schema and Prototype, The International Journal of Construction Information Technology, Vol. 2, No. 4, pp. 63–80.

Turk, Ž., Björk, B.-C., Johansson, C. and Svensson, K. (1994) Document Management Systems as an Essential Step Towards CIC, Preproc CIB W78 workshop on Computer Integrated Construction, VTT, Helsinki, Finland, 22–24 August.

Wager, D. and Winterkorn, E. (1998) Document Management for Construction, CICA, UK, ISBN 0–906225–23–1, pp. 182.

Yogeswaran, K. and Kumaraswami, M. M. (1997) Rejuvenating Contract Documentation to Reflect Realistic Risk Allocations, Proceedings of the CIB W78 Workshop, IT Support for Construction Process Re-Engineering, Cairns, Australia, 9–11 July, pp. 433–442.

Zarli, A. and Rezgui, Y. (2000) A Survey of Internet-oriented Technologies for Document-driven Applications in Construction Open Dynamic Virtual Environments, Proceedings of Construction IT 2000, CIB W78, IABSE, EG-SEA-AI, ISBN 9979–9174–3–1, Reykjavik, Iceland, 28–30 June, pp. 1089–1101.

Chapter 10

Enabling Concurrent Engineering through 4D CAD

Sheryl Staub-French and Martin Fischer

10.1 Project overview

To illustrate how to apply 4D models for project planning, this chapter presents lessons learned from the application of 4D models on a bio-technology project. The application of 4D models was part of a larger effort to investigate, through the live application of an integrated suite of project design and management software tools on a design-build project, how an integrated project team can utilise 3D CAD models linked to cost estimates and schedules to design, coordinate, estimate, plan, schedule and manage a construction project. We investigated the capabilities of existing software tools to leverage 3D models for design coordination and constructability analysis, cost estimating, and construction planning. Our insights are based on our work with the project team during design and construction of the 'Sequus Pharmaceuticals Pilot Plant'. Each team member committed to modeling their respective scope of work in 3D CAD from the beginning of design through construction completion, coordinating the designs in 3D, and using commercial software to integrate the 3D CAD models with cost estimating and scheduling software. Through the case study, we determined the benefits and shortcomings of this suite of project planning and management software, the specific steps required to accomplish integration of scope, cost and schedule information and lessons learned with respect to 4D modeling and the impact of these tools on project performance. This chapter shows that early and simultaneous involvement of a project team including designers, general contractors and subcontractors in the design and construction of a capital facility coupled with the use of shared 3D and 4D models allows the project team to deliver a superior facility in less time, at lower cost and with less hassle. Compared to the traditional, sequential and paper-based design, planning and construction process, the construction project team realised the following specific benefits during design and construction of the Pilot Plant Facility for Sequus Pharmaceuticals:

- shorter estimating time,
- fewer quantity takeoff errors,

- better documentation and reproducibility of the estimating process,
- elimination of field interferences,
- improved communication of the schedule intent,
- construction completed on time and under budget,
- less rework,
- increased productivity,
- 60 per cent fewer requests for information,
- fewer change orders,
- less than 1 per cent cost growth and
- decrease in time from start of construction to facility turnover.

While this chapter focuses on the application and benefits of 4D modelling for detailed planning and coordination of construction, we also investigated the usefulness of existing design-cost integration software and identified necessary research to meet the practical needs of cost estimators (Staub-French and Fischer, 2001). Our role on the construction project was to assist the project professionals with the generation of 3D and 4D information, to help them integrate the scope and schedule information, and to document steps, benefits, barriers and research needs for the use of concurrent design tools.

10.2   Construction project and technology overview

The project's scope was to construct a pilot plant facility within an existing warehouse for Sequus Pharmaceuticals, a bio-tech company located in Menlo Park, California. Figure 10.1 shows the 2D view of the 3D architectural model developed by the architect, Flad & Associates. The facility contains 20,000 square feet of available space, with 3,440 square feet of office space, 3,100 square feet of manufacturing space, 2,900 square feet of process development space, and 4,800 square feet of future expansion space. The project started construction in May 1998 and construction was substantially complete as scheduled on 1 February, 1999. The negotiated contract price was approximately US$5,800,000.

The Sequus project was unique in that the core project team consisting of the design firms, general contractor and three key subcontractors focusing on mechanical, electrical, and piping work was assembled prior to design and construction, with each team member committed to modeling their respective scope of work in 3D CAD using a design-build, concurrent engineering (CE) approach. The general contractor assembled the design-build team based on each company's experience using 3D CAD technology on past construction projects and previous experience working with each other. The design firms were responsible for providing the basis of design and schematic drawings for the mechanical, electrical and piping (MEP)

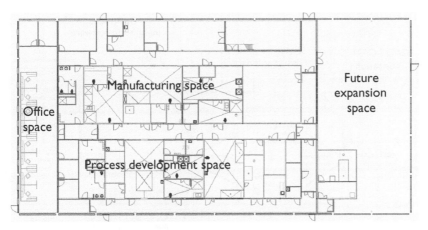

Office space

Manufacturing space

Future expansion space

Process development space

Figure 10.1 Architectural layout of Sequus project.

work while the subcontractors were responsible for the detailed design and 3D modelling of their scope of work. The general contractor was responsible for orchestrating and managing the distribution of electronic design information. The project team consisted of the following companies: the design firm Flad & Associates, the General Contractor Hathaway/ Dinwiddie Construction Company (Hathaway), the engineering firm Affiliated Engineers Incorporated, the mechanical subcontractor Paragon Mechanical, the electrical subcontractor Rosendin Electric, and the piping subcontractor Rountree Plumbing & Heating.

A primary goal of the project was to leverage the detailed 3D model of the facility throughout design and construction and explore the use of existing software to integrate the 3D CAD models with cost and scheduling software (Staub *et al.*, 1999). Designing in a collaborative environment and utilising design, cost and schedule integration software forced and enabled the project team to work together from the very start of design and to share information throughout all phases of design development. Figure 10.2 shows the software used by each discipline to create the 3D CAD models and the software used to integrate design, cost, and schedule information. Consequently, all members of the project team were able to use the software tools with which they were familiar, which were tools that leveraged the 3D design information for their work. For example, the piping subcontractor used the electronic 3D model in Multi-Pipe for fabrication. The CAD-estimate link developed by Ketiv and Timberline supported design-cost integration and Bentley's Schedule Simulator provided design and schedule integration (4D). At the time of this project the link between Ketiv's 3D and Timberline's cost estimating software was the only commercially available,

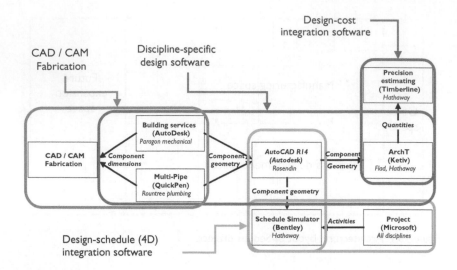

Figure 10.2 Software used to perform design, cost and schedule integration.

off-the-shelf integration of scope and cost estimating information available. With the implementation and certification of information exchange mechanisms based in the industry foundation classes (IFC) standard (Yu *et al.*, 1998) in 2001 in several software applications there is now more flexibility in selecting 3D design tools and linking them to cost estimating tools like Timberline's Precision Estimating.

Traditional construction planning tools, such as bar charts and network diagrams, do not represent and communicate the spatial and temporal, or four-dimensional, aspects of construction schedules effectively. Consequently, they do not allow project managers to create schedule alternatives rapidly to find the best way to build a particular design. Extending the traditional planning tools, visual 4D models combine 3D CAD models with construction activities to display the progression of construction over time. 4D models combine 3D CAD models with the project timeline. Systems linking 3D CAD models with schedule and other project information started to be developed in the mid-eighties (Kahan and Madrid, 1987; Atkins, 1988). Cleveland (1989) reported on the development and application of 4D models in the R&D group at Bechtel. The 4D tool development efforts at Bechtel were part of a larger effort to enhance project management through data integration and visualisation. The Bechtel R&D effort eventually led to the start of Jacobus Technology and the marketing of the Construction Simulation Toolkit (CST). See, for example (Collier and Fischer, 1996) for a discussion of the use of CST on a project. CST evolved into what is now known as the

Bentley Schedule Simulator (the 4D software we used on the Sequus project). Separately, Bechtel further developed 4D technology internally (Williams, 1996). In the 90s, experience on many different types of projects (simple to complex, new to retrofit) has shown that combining scope and schedule information in one visual model is a powerful communication and collaboration tool for technical and non-technical stakeholders (Retik, 1997; Edwards and Bing, 1999; EPRI, 2000; Schwegler et al., 2000; Haymaker and Fischer, 2001).

10.3   Goals and preparation for 4D modelling on the Sequus project

Bio-tech facilities typically have very complex MEP systems. Consequently, the challenging aspects of this project were coordinating the design of the different yet interdependent MEP systems and installing the complex MEP systems within the confined space in the existing warehouse facility. The design in 3D of these systems and the collaborative and team-oriented approach ensured that the designs were coordinated and that conflicts were avoided. The MEP design coordination process was accomplished through weekly coordination meetings and continuous information sharing using a project FTP site. During the design coordination meetings, the CAD modelers for each subcontractor would huddle around a large computer screen and inspect the electronically integrated CAD models. By integrating each discipline's scope of work in 3D, each discipline could better visualise their relationships to other trades, identify design conflicts easily, and explore alternative solutions in a 3D space. This allowed the project team to identify and eliminate most design conflicts prior to the start of construction. The only design conflicts identified in the field were between trades that did not model their work in 3D. There was only one documented field interference among the MEP contractors.

The MEP systems were designed such that the majority of the work was placed on an equipment platform. The platform was necessary because the existing structure was not capable of supporting the increased loads from the MEP systems and related equipment. The distribution of the MEP systems in the interstitial space between the platform and the first floor ceiling was a challenging task. Figure 10.3 shows the MEP systems and related equipment on the equipment platform. The late arrival of the Air Handler Units (AHU's) further complicated the installation of these systems. The AHU's were not scheduled for arrival until one month after the MEP system installation had started. Consequently, a goal of the coordination process was to limit the interaction between the subcontractors installing the different systems and to keep the installation path for the AHU's open. The installation of the MEP systems and equipment installation was modeled in 4D to ensure that it could be executed effectively. A later section

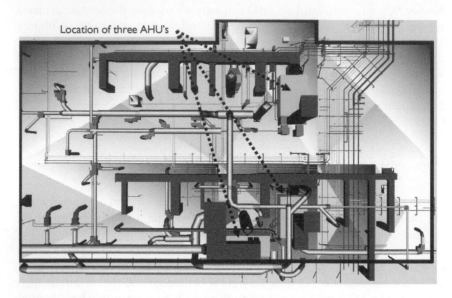

Figure 10.3 Top-down view of equipment platform.

will discuss the use and benefits of the detailed 4D model of the equipment platform.

10.4   Scope of 4D modelling effort

The complexity of the MEP systems and late arrival of the AHU's heightened the demand for coordination on the equipment platform. Consequently, a 4D model of all work on the equipment platform was constructed to ensure that interferences between trades were avoided, rework minimised, and productivity maximised. The 4D model included the following scope of work: mechanical, electrical, process piping, decking and equipment located on the equipment platform.

The 3D model of the Sequus Pilot Plant Facility contained 32,753 objects, with approximately 17,900 3D objects representing the equipment platform. Table 10.1 shows the number of 3D objects for each system. The final schedule contained over 1,300 activities including pre-construction, major procurement items and construction. The schedule contained approximately 55 activities to coordinate the MEP and related equipment installation on the equipment platform.

The purpose of the 4D model was to assist with the coordination of the subcontractors installing the MEP work on the equipment platform. Consequently, the general contractor and the subcontractors constructed the 4D model to show the day-to-day completion of the MEP system

Table 10.1 Number of 3D objects generated to represent MEP systems

Company	System	Description of work		# of 3D Objects
Rountree	Process piping Mechanical – wet side	Equipment room		2,915
		Compressed/instrument air		2,541
		Hot water/cold water		7,369
		Chilled water		4,616
		Utilities		5,452
		Water for injection (WFI)		873
			Sub total	23,766
Paragon	Mechanical – dry side	Distribution below platform		3,311
		Distribution and equipment above platform		2,397
			Sub total	5,708
Rosendin	Electrical	Power/lighting		760
Flad	Architectural	Architectural		2,519
Hathaway	Structural	Equipment platform		355
Total				32,753

installation and workflow on the equipment platform. Specifically, the 4D model graphically illustrated the following:

- *where* subcontractors could and could not work on any given day,
- the areas that were not *available* until after AHU's were installed,
- the areas where *multiple trades* were working in close proximity simultaneously, and
- the *feasibility* of the planned delivery of AHU's to meet owner's approval.

Therefore, the 4D model represented the subcontractor's perspective and assisted with the day-to-day coordination of their work and the related equipment installation.

The purpose of the 4D model for research was to identify the functionality needed to assist with subcontractor coordination of day-to-day construction operations. The 4D models created in previous research had primarily focused on the owner's or general contractor's perspective and communicated schedule intent but not schedule detail (Collier and Fischer, 1996; Koo and Fischer, 1998). Although the 4D model was used for that purpose on the Sequus project also, the primary purpose was to explore the usefulness of 4D models to represent the subcontractor's perspective.

10.5   Construction planning process with 4D

We worked with the project team throughout design and construction to assist with the collaborative design and construction processes. Throughout the pre-construction phase, we helped the subcontractors and architect organise their 3D models to facilitate the MEP design coordination process and the integration of the 3D models with cost and schedule (4D) integration software. Throughout construction, we primarily worked with the general contractor and subcontractors to assist with the coordination of MEP work on the equipment platform using the 4D model.

Creating the 4D model was a three-step process: (1) elaborate the schedule, (2) group the 3D objects and (3) create the 4D model. Figure 10.4 illustrates the 4D model generation process.

10.5.1   Step 1 elaborate schedule

First, together with the subcontractors, we expanded the master schedule created by the general contractor to the level of detail required to represent the day-to-day operations of the various subcontractors. We consulted the foreman for each of the three MEP trades and the superintendent for the general contractor to determine what activities were necessary and how work would flow through the equipment platform. We then added the necessary detail to the schedule and divided the activities for the MEP work into seven zones to represent workflow. Consequently, the schedule showed when each of the subcontractors would be working in each zone on the equipment platform. The master schedule started with 10 activities for the installation of the MEP work on the equipment platform. The elaboration step took approximately eight hours and added 45 activities to the master schedule. Figure 10.5 shows the activities added to the schedule for the piping installation.

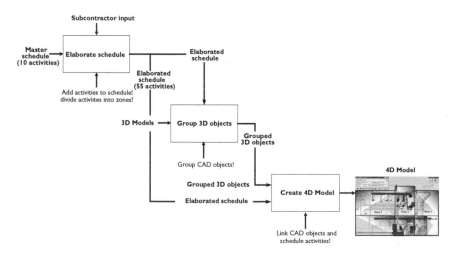

Figure 10.4 IDEF0 diagram of 4D model generation process.

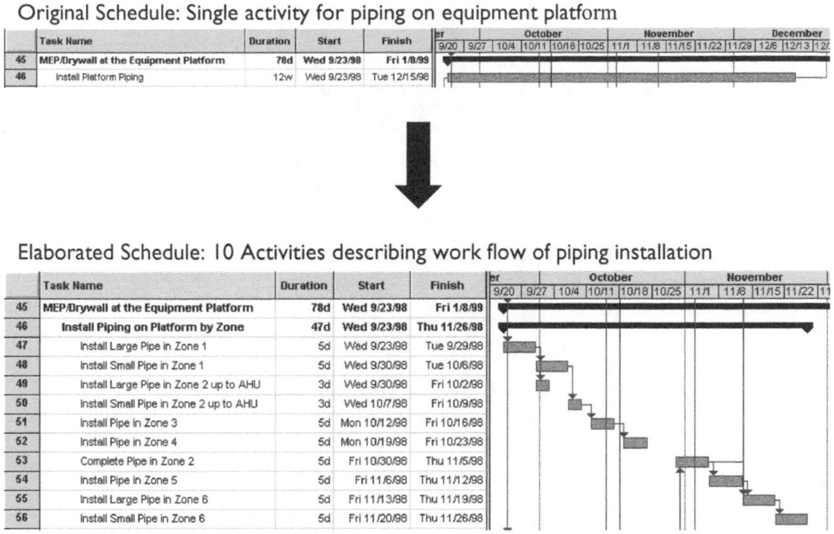

Original Schedule: Single activity for piping on equipment platform

	Task Name	Duration	Start	Finish	ber 9/20	9/27	October 10/4	10/11	10/18	10/25	11/1	November 11/8	11/15	11/22	11/29	December 12/6	12/13	12/
45	MEP/Drywall at the Equipment Platform	78d	Wed 9/23/98	Fri 1/8/99														
46	Install Platform Piping	12w	Wed 9/23/98	Tue 12/15/98														

Elaborated Schedule: 10 Activities describing work flow of piping installation

	Task Name	Duration	Start	Finish	ber 9/20	9/27	October 10/4	10/11	10/18	10/25	11/1	November 11/8	11/15	11/22	11
45	MEP/Drywall at the Equipment Platform	78d	Wed 9/23/98	Fri 1/8/99											
46	Install Piping on Platform by Zone	47d	Wed 9/23/98	Thu 11/26/98											
47	Install Large Pipe in Zone 1	5d	Wed 9/23/98	Tue 9/29/98											
48	Install Small Pipe in Zone 1	5d	Wed 9/30/98	Tue 10/6/98											
49	Install Large Pipe in Zone 2 up to AHU	3d	Wed 9/30/98	Fri 10/2/98											
50	Install Small Pipe in Zone 2 up to AHU	3d	Wed 10/7/98	Fri 10/9/98											
51	Install Pipe in Zone 3	5d	Mon 10/12/98	Fri 10/16/98											
52	Install Pipe in Zone 4	5d	Mon 10/19/98	Fri 10/23/98											
53	Complete Pipe in Zone 2	5d	Fri 10/30/98	Thu 11/5/98											
54	Install Pipe in Zone 5	5d	Fri 11/6/98	Thu 11/12/98											
55	Install Large Pipe in Zone 6	5d	Fri 11/13/98	Thu 11/19/98											
56	Install Small Pipe in Zone 6	5d	Fri 11/20/98	Thu 11/26/98											

Figure 10.5 Original activity and elaborated activities for piping installation.

10.5.2   Step 2 group 3D objects

We used the 3D models created by the architect and MEP subcontractors to create the 4D model. However, the 3D models represented the designers' perspective and needed to be organised differently to represent the construction perspective. Essentially, each layer in the 3D model needed to be organised so that it corresponded to an activity in the schedule. Consequently, we created new layers, renamed old layers, and moved CAD objects to the appropriate layer. For example, in the electrical 3D drawing, there were two separate layers for wiring for lighting and wiring for power. For scheduling purposes, it was necessary to distinguish wiring by whether it was in the ceiling or in the wall. Therefore, the corresponding layers and objects had to be changed to 'wall rough-in' and 'ceiling rough-in'. In addition, the 3D CAD models also had to be reorganized to incorporate the workflow through the equipment platform. Consequently, the 3D CAD models had to be reorganised so that the scope of work related to each of the seven zones was assigned to a separate layer. To illustrate the extent of changes required for this step, the HVAC design model originally contained six layers. After we had modified the model to correspond to the schedule activities, there were 22 layers. The piping design was particularly cumbersome to organize because each of the different piping systems were contained in separate drawing files. Figure 10.6a shows the original piping drawings organized by piping system, and Figure 10.6b shows the revised

(a)

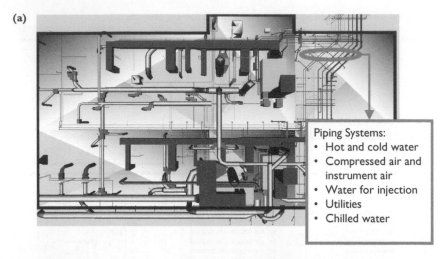

Piping Systems:
- Hot and cold water
- Compressed air and instrument air
- Water for injection
- Utilities
- Chilled water

(b)

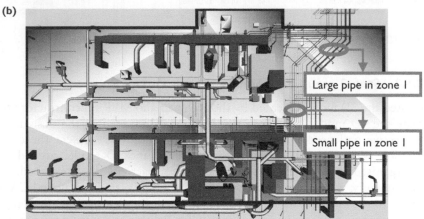

Large pipe in zone 1

Small pipe in zone 1

Figure 10.6 Original and reorganised piping drawings of equipment platform: (a) Original piping drawings organised by piping system; (b) Revised piping drawings organised by workflow.

piping drawings organised by workflow. We performed this process on five piping drawings, the HVAC drawing for the ductwork and AHU's, and the structural drawing containing the concrete decking. The total duration for this task was 16 hours.

10.5.3   Step 3 create 4D model

To create the 4D model, we used Bentley's Schedule Simulator. This software imports 3D CAD models and schedule models and transforms them

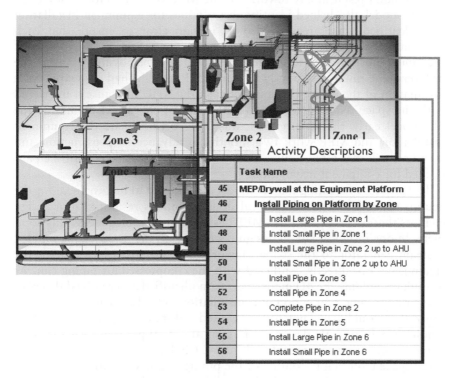

Figure 10.7 Grouped CAD objects and schedule activities for piping in zone 1. Reorganised design: organised by zones.

into object-oriented models. We imported each of the 3D CAD models as separate files so that we could easily focus on specific systems in the 4D models. Consequently, we imported eight 3D CAD files into the Schedule Simulator, which allowed us to view any combination of the different systems in 4D. After the CAD models and schedule model was imported, we manually related the grouped 3D CAD objects created in the second step with the appropriate schedule activity created in the first step. Figure 10.7 shows the grouped CAD objects for large and small piping in zone 1 and the corresponding schedule activities. This step took approximately four hours.

10.6 Use of 4D model

The 4D model was primarily used as a communication tool between the general contractor and the owner and between the general contractor and the subcontractors. The 4D model of the work on the equipment platform demonstrated to the owner that the equipment could be installed as planned and wouldn't result in any rework for the MEP subcontractors. Moreover, the 4D model also helped identify access issues for equipment installation and identified what areas needed to remain clear to ensure that each subcontractor could install equipment as planned. Specifically, it showed the piping subcontractor foreman that it would not be possible to install the different pipe runs continuously as planned. Rather, he had to postpone the installation of the piping that ran between the AHU's because it interfered with the space required for the AHU's installation path. By building the 4D model early, the project team was able to coordinate the equipment installation and MEP work two months in advance of construction and avoid rework that often results when work in place conflicts with the path needed for equipment installation.

The process of creating the 4D model also proved to be beneficial for the Sequus team. We identified several design conflicts resulting from design changes that occurred after the MEP design coordination process was complete. In one instance, late in design development, the designers had added a steam generator to the scope of work. The proposed location for the steam generator directly conflicted with the compressed air piping run. As a result of building the 4D model, the team identified and resolved this conflict prior to pipe fabrication and installation. We also identified a design error that could have potentially caused substantial rework. The AEC Chiller was incorrectly designed in 3D at about 20 per cent its actual size. After correction of this mistake, the AEC Chiller no longer fit in the allocated space requiring re-routing of the piping to a new location. Thanks to constructing the project first in 3D and 4D with a CE approach, the team also identified and resolved this conflict months before the AEC Chiller was scheduled for installation.

Construction of the MEP equipment and system proceeded seamlessly in the field. The piping and mechanical subcontractors reported increased field productivity, less rework, and fewer change orders and requests for information than expected for a job of this complexity, as shown in Table 10.2. For the scope of work modeled in 3D, there was only one contractor-initiated change order, which is remarkable for work of this complexity. There were about 60 per cent fewer requests for information (RFIs) on this project compared to a project using a traditional and less integrated process. These advantages were somewhat offset by a longer design and detailing process.

10.7   Lessons learned with respect to 4D functionality

The *visualisation* and *communication* capabilities of the 4D tool were the most useful functionality. The 4D model was very detailed and showed the day-to-day operations of all the MEP subcontractors. The 4D tool was used by the general contractor to visualise the MEP subcontractors' workflow on

Table 10.2 Summary of MEP subcontractors' experience

	Rountree Plumbing Process Piping/HVAC Wet	Paragon mechanical HVAC dry	Rosendin electric Electrical
Contract value	$2,018,937	$1,071,237	$488,414
Increased design costs and time	30%	20–30%	300%
	Difficult to find trained designers with installation experience File size with Solid pipe designer	Went from 2D to 3D	Engineering costs typically 4% engineering costs – 12%
Number of change orders	6	1	3
Reason for change orders	4 – Owner requested 1 – Value engineering 1 – Unforeseen condition	1 – Owner requested	3 – Owner requested
Change in project costs due to change orders	–1.00%	1.00%	0.97%
Number of RFI's	40	23	—
Expected RFI's	100	50	—
Example conflict avoided	Routing of chilled water and heating water to AHU	Relocated reheat coils to avoid ductwork conflict	Coordination or reflected ceiling plan and register location
Productivity	Significantly increased	Much more productive	No difference
Rework	Dramatically reduced – Only occurred on non-3rd portions	Minimal	No difference
Profitability	Same Expects greater return with increased use	Same Expects greater return with increased use	Less Increased design time with less benefit from coordination when compared to other trades

the equipment platform and to assess the installation path for the AHU's so rework could be avoided and productivity maximised. The general contractor also used the 4D model to visually communicate the schedule to the owner to demonstrate the feasibility of the proposed schedule. The general contractor then used the 4D tool to communicate to the MEP subcontractors when and where they could work and to alert them to the areas that must be left clear to leave space for later equipment installation. The team also found it useful to visualise the planned status of construction at any point during construction.

There are many opportunities to improve the capabilities of existing 4D tools. Our recommendations for improvement based on this case study are as follows:

- Provide mechanisms that transform the design model into a model that represents the construction perspective (Fischer *et al.*, 1998). Today, planners need to modify the CAD model so that the CAD objects can be related to the associated schedule activities. For example, the electrical design model had two layers for electrical and power wiring. To represent the construction perspective, the objects on these layers had to be changed into 'wall rough-in' and 'ceiling rough-in' to correspond with the installation process. This is a time-consuming task that has to be repeated if the design changes. Such mechanisms are particularly important for situations when the design changes frequently because the manual process of reorganising the 3D CAD objects does not establish a dynamic link between the designers' and the constructors' organisation of the 3D objects. Hence, without such mechanisms, the reorganisation has to be repeated whenever the design changes.
- Provide mechanisms that automatically transform the design models to correspond with workflow changes (Akbas *et al.*, 2001). During the planning of the scope of work on the equipment platform, the general contractor decided to break the platform into seven zones with work flowing counterclockwise. As a result, we manually changed all the CAD models to correspond with this plan. However, this may not have been the optimal workflow. If this transformation could be automated, project teams could explore a variety of installation plans to optimise the installation process.
- Provide functionality that allows project teams to create the schedule right in the 4D system (McKinney *et al.*, 1996; Fröhlich *et al.*, 1997). Today, a schedule needs to exist before a 4D model can be built. Consequently, project teams must wait for the completion of the 4D model before they can visualise their planning decisions.
- Provide analysis tools to help project teams understand the impact of their design and planning decisions (Akinci *et al.*, 2000). Today, analysis of 4D models is performed manually. For example, we identified the

conflict between the pipe runs and the installation path for the AHU's through visual inspection of the 4D model. It would be very useful if the computer could identify such constructability problems automatically.

• Explicitly link the 4D model(s) to the 3D model(s) (Fischer and Aalami, 1996). 4D models are created by importing 3D models into a 4D modeling tool (Bentley's Schedule Simulator in our case). Consequently, if the design changes, the 4D model has to be manually created again.

10.8   Lessons learned with respect to 4D process and organisation

This case study suggests that early and simultaneous involvement of designers, general contractors, and subcontractors in the design and construction of facilities coupled with the use of shared 3D and 4D models allows AEC (architecture, engineering, construction etc.) service providers to deliver a superior facility more reliably in less time, at lower cost, and with fewer hassles. Moreover, to leverage the 3D models throughout design and construction, it is important that the organisations that have the most to benefit from the 3D models actually create them (Fischer, 2000). For example, the subcontractors created the detailed 3D models for the complex MEP systems that were utilised for electronic design coordination, daily coordination during construction using 4D models and automatic fabrication.

This case study also suggests that owners, designers, and builders of facilities will need to develop new skills and implement organisational changes to capitalise on the benefits offered by this technology. Specifically, owners will need to bring a project team together early in the project. Designers will need to focus more on the overall design and coordination of design tasks and less on detailed design. General contractors will need to learn how to manipulate 3D CAD models, work more closely with the designers during design development, and provide input on how to model designs in 3D so that the CAD models are more usable by constructors. Finally, subcontractors will also need to learn design software, as they will be performing more detailed design, working more closely with the architects and engineers through the design process, and addressing coordination issues early in design development.

10.9   Acknowledgements

This work was supported by the National Science Foundation under Grant. It was also supported with matching funds from the following firms: Flad & Associates, Hathaway-Dinwiddie Construction Company, Mazzetti and Associates, Paragon Mechanical, Rountree Plumbing, and Rosendin

Electric. We also thank the following software firms for their support throughout this project: Bentley, Ketiv, Timberline.

10.10   References

Akbas, Ragip, Fischer, Martin, Kunz, John and Schwegler, Benedict (2001). 'Formalizing Domain Knowledge for Construction Zone Generation'. Proceedings of the CIB-W78 International Conference IT in Construction in Africa 2001: Implementing the next generation technologies, May 30–June 1, CSIR, Division of Building and Construction Technology, Pretoria, South Africa, pp. 30-1 to 30-16. Available at http://buildnet.csir.co.za/constructitafrica/authors/Papers/accepted.htm

Akinci, Burcu and Fischer, Martin (2000). '4D WorkPlanner – A Prototype System for Automated Generation of Construction Spaces and Analysis of Time-Space Conflicts'. Eighth International Conference on Computing in Civil and Building Engineering (ICCCBE-VIII), Renate Fruchter, Feniosky, Pena-Mora and W. M. Kim Roddis (eds), ASCE, Reston, VA, pp. 740–747.

Atkins, David C. (1988). 'Animation/Simulation for Construction Planning'. Engineering, Construction, and Operations in Space: Proceedings of Space 88, ASCE, 670–678.

Cleveland, A. B., Jr (1989). 'Real-Time Animation of Construction Activities'. Proceedings of Construction Congress I – Excellence in the Constructed Project, ASCE, New York, 238–243.

Collier, Eric and Fischer, Martin (1996). 'Visual-Based Scheduling: 4D Modeling on the San Mateo County Health Center'. Proceedings of the Third Congress on Computing in Civil Engineering, Jorge Vanegas and Paul Chinowsky (eds.), ASCE, Anaheim, CA, June 17–19, 1996, 800–805.

Edwards, Ross and Zeng, Bing (1999). 'Case Study: 4D Modeling & Simulation for the Modernization of Logan International Airport'. Proceedings of the 1997 International Conference on Airport Modeling and Simulation, ASCE, 8–27.

EPRI (2000). 'Virtual Reality Construction: 4-D Visualization', EPRI, Palo Alto, CA, and CE Nuclear Power LLC Windsor, CT.

Fischer, Martin (2000). 'Benefits of 4D Models for Facility Owners and AEC Service Providers'. 6th Construction Congress, ASCE, Kenneth D. Walsh (ed.), Orlando, FL, Feb. 20–22, 2000, 990–995.

Fischer, Martin A. and Aalami, Florian (1996). 'Scheduling with Computer-Interpretable Construction Method Models'. Journal of Construction Engineering and Management, ASCE, 122(4), 337–347.

Fischer, M., Aalami, F. and Akbas, R. (1998). 'Formalizing Product Model Transformations: Case Examples and Applications'. Artificial Intelligence in Structural Engineering: Information Technology for Design, Collaboration, Maintenance, and Monitoring, Springer Verlag, 113–132.

Fröhlich, Bernd, Fischer, Martin, Agrawala, Maneesh, Beers, Andrew and Hanrahan, Pat (1997). 'Collaborative Production Modeling and Planning', Computer Graphics and Applications, IEEE, 17(4), 13–15.

Haymaker, J. and Fischer, M. (2001). 'Challenges and Benefits of 4D Modeling on the Walt Disney Concert Hall Project'. Working Paper 064, Center for

Integrated Facility Engineering, Stanford University, Stanford, CA. Available at: http://www.stanford.edu/group/CIFE/Publications/index.html (accessed July 2006).

Kahan, E. T. and Madrid, X. H. (1987). 'Integrated System to Support Plant Operations'. *Hydrocarbon Processing Symposium*, ASME, 55–60.

Koo, B. and Fischer, M. (1998). 'Feasibility Study of 4D CAD in Commercial Construction', *Journal of Construction Engineering and Management*, ASCE, 126(4), 251–260.

McKinney, K., Kim, J., Fischer, M. and Howard, C. (1996). 'Interactive 4D-CAD'. *Proceedings of the Third Congress on Computing in Civil Engineering*, Jorge Vanegas and Paul Chinowsky (eds), ASCE, Anaheim, CA, June 17–19, 383–389.

Retik, A. (1997). 'Planning and Monitoring of Construction Projects Using Virtual Reality'. *Project Management*, 3(1), 28–31.

Schwegler, Ben, Fischer, Martin and Liston Kathleen. (2000). 'New Information Technology Tools Enable Productivity Improvements'. North American Steel Construction Conference, American Institute of Steel Construction (AISC), Las Vegas NV, February. 23–26, pages 11–1–11–20.

Staub, S., Fischer, M. and Spradlin, M. (1999). 'Into the Fourth Dimension'. *Civil Engineering*, ASCE, 69(5), 44–47.

Staub-French, Sheryl and Fischer, Martin (2001). 'Industrial Case Study of Electronic Design, Cost, and Schedule Integration'. *Technical Report Nr. 122*, Center for Integrated Facility Engineering (CIFE), Stanford University, Stanford, CA, accessed at http://www.stanford.edu/group/CIFE/Publications/index.html on June 1, 2001.

Williams, M. (1996). 'Graphical Simulation for Project Planning: 4D-Planner', *Proceedings of Third Congress on Computing in Civil Engineering*, ASCE, New York, 404–409.

Yu, Kevin, Froese, Thomas and Grobler, Francois (1998). 'International Alliance for Interoperability: IFCs'. *Proceedings of Congress on Computing in Civil Engineering*, ASCE, 395–406.

Chapter 11

Telepresence Environment for concurrent lifecycle design and construction

Chimay J. Anumba and A. K. Duke

11.1 Introduction

The adoption of Concurrent Engineering (CE) principles in the construction industry will increase the level of collaboration between the parties involved in projects. One aim of CE is to address downstream issues early on in the project life-cycle. This calls for the involvement of all parties (including specialist subcontractors) at a much earlier stage than would be the case in a traditional construction project environment. It is not feasible for all these parties to be co-located during this period of design and hence the reliance of the project group upon information and communications technologies will increase. The deployment of these technologies throughout the project lifecycle will be beneficial for effective collaboration.

Within a CE setting, there is the need for an integrated information and collaboration environment that will create a persistent space to support interaction between project personnel throughout all phases of construction projects. A communication architecture that will provide an infrastructure for such an environment is the main focus of this paper. The user's perspective is defined and the modes of interaction are explained using the architecture as a basis. The primary features of the environment are visual representations of the construction project and of the people working on it. These representations provide access to project information via integration with project data management systems and access to the people via integrated communication channels. As well as acting as a unified interface for the project and its people, this is intended to improve collaboration by promoting serendipitous contact. Agent technology is employed to allow the environment to draw users towards people with common issues or interests and towards elements of the project information that are of potential interest. The intention is to recreate for remote people the ad hoc meetings and informal cues that are so important for collaboration when people are co-located.

This chapter first identifies the key communications issues that need to be addressed within a 'Concurrent Lifecycle Design and Construction' (CLDC) environment. It then discusses the concept of Telepresence and, using examples from an initial research prototype, demonstrates how it can support collaborative and CE in construction. The outline features of an advanced Telepresence environment for construction, which builds on the earlier research prototype described, are also presented.

11.2   Communications issues in CLDC

The key communications issues in CLDC reflect the changing environment within which the construction industry operates. Some of the key aspects of this environment include the following (Anumba and Evbuomwan, 1999):

- Concurrency in an integrated design and construction process requires greater discipline in the production, manipulation, storage and communication of project information.
- Project information necessarily consists of both graphical and non-graphical information, which must be communicated between members of the project team.
- The greater the level of concurrency in a process, the greater the level of co-ordination required. This entails an increased level of communication between the various stages and activities in the process, as well as between the project team members.
- Paper-based communication of project information is now inadequate to cope with the high level of functionality (in terms of speed, accuracy, usability, ease of modification, enhanced visualisation, improved co-ordination, etc.) required in a collaborative working environment.
- The increasing 'globalisation' and complexity of construction projects means that project teams often involve partners from widely distributed geographical areas, sometimes on different continents (Madigan, 1993). Face-to-face meetings in such circumstances are expensive in terms of time, money and personal inconvenience (Rogers, 1994); effective communication protocols able to collapse time and distance constraints are therefore necessary.
- The very fast pace of technological development, particularly in computing and telecommunication dictate that, for the construction industry to remain competitive, it must take advantage of new and emerging information and communication technologies such as the internet, multimedia, virtual reality, broadband communication networks, etc.

In the light of the above, there is the need for a clear identification of distinct groups of people, tools and project phases across which communication has to take place. There are seven main facets of communication that need to be addressed in CLDC. These have been discussed in detail by Anumba and Evbuomwan (1999) and include:

- communication between intra-disciplinary CAE tools (F1);
- communication between each project team member and his/her design tools (F2);
- communication between project team members (F3);
- communication between each discipline and the common project model (F4);
- communication across the stages in the project life-cycle (F5);
- communication between the project team and third parties (F6);
- communication between inter-disciplinary CAE tools (F7).

Various technologies are now being employed to address the above communication issues and new ones are emerging. One technology that is expected to have major impact, particularly on communication between project team members, is Telepresence.

11.3   The concept of Telepresence

Telepresence may be defined as 'the ability to operate a device by remote control, including perceptual data and sensory feedback transmitted from the operator, such that it appears to the operator as if the operator were present at the site of the remote device and operating it directly' (Morris, 1992). This is a rather broad definition and it is important to define more clearly the context within which the term is used in this thesis. Within a collaborative communications setting, Telepresence can be viewed as the facility which enables collaborating parties to be virtually co-located within a given (3D) environment, in which they are able to interact with one another or with virtual objects that are also present in that environment (Anumba and Duke, 1997). The intended aim of this being to create the illusion of 'being there' (Cochrane *et al.*, 1993). This is perhaps Telepresence in its purest sense. Another definition – 'Telepresence is enabling human interaction at a distance, creating a sense of being present at a remote location' (Walker and Sheppard, 1997) – implies that technologies such as the telephone (by extending human speech and hearing) or video conferencing (by also extending vision) provide Telepresence to a degree. The other end of the spectrum is embodied by technologies such as the VisionDome™[1] (Traill *et al.*, 1997). This immersive projected display technology (shown in Figure 11.1) can be used to provide a high degree of Telepresence.

It is evident from the above definitions that Telepresence systems have significant potential for improving communications in a variety of settings. Equipment maintenance/installation, mobile news-gathering, telemedicine, and remote surveillance are just a few of the emerging applications

Figure 11.1 Architectural review meeting inside the VisionDome.

(Cochrane *et al.*, 1993). There is major scope for enhancing construction project team communications through the use of Telepresence. Duke, Bowskill and Anumba (1998) have identified specific areas where Telepresence could be of use in a concurrent life-cycle design and construction setting. These include facilitating multi-disciplinary teams, integrating communications facilities with design tools and supporting project team communications with the use of collaborative virtual environments.

11.4 A Telepresence Environment for construction

The principal aim of the Environment is to support CLDC projects by providing a collaborative space for personnel that integrates access to people and information. The Environment contains visual representations of the construction project and of project team members (i.e. the users). These representations provide the users with access to underlying project data (such as drawings, schedules, rationale, etc.) and to the other users via integrated communication channels.

The Environment aims to provide support to users in two ways:

- Passive – to allow people to maintain an awareness of others and of the construction project. The environment draws the individual towards features and people that are of interest or are currently impacting upon their work area. The aim is to facilitate serendipitous contact with people and information via interest profiling.
- Active – to act as a ubiquitous user-interface for finding, contacting and communicating with people and for locating information about the construction project, thus providing teleconferencing with a context.

In both scenarios, the aim is allow project personnel to work as if co-located with colleagues and as if located at the construction site.

The technological basis for the system is Internet Protocol (IP) or, more specifically, Web technology. This allows users to access information stored in the network using a common client or browser. Thus, it is essential that all project information should be accessible over a project extranet and that existing project systems should be Web-enabled.

The usage of the system will be persistent that is a client is started as a user logs-on and remains running throughout the day or working session. The contextual load placed on the user is variable and adapts to different usage situations. The load is varied by techniques such as changing the window size of the client or altering the style of information delivery. For example, when users are performing activities such as writing a document, the window size reduces giving users a 'background' awareness of the information being presented. Upon seeing something of importance, users can select the client and increase the window size. The window can also be increased to full screen during periods of inactivity – rather like a screensaver.

Integral to the client is a 3D browser. This enables 3D environments to be downloaded and displayed and allows users to navigate around it. The environment contains a 3D representation of the construction project. It is multi-user and contains avatars to represent the other users. Objects and avatars in the environment contain hotlinks to information. In the case of avatars, these are links to the corresponding user's homepage. Construction objects have links to project documents (drawings, schedules, rationale etc.). As stated above, the system is dependent upon a single project repository or a Web-enabled document management system. By associating URLs with objects in the environment, access to stored information is given. Since this information might only be available to specific people a user authentication process is required. This is carried out as users join the environment and then used whenever restricted documents are accessed.

11.4.1   *Project information integration*

As stated above, the environment relies upon existing systems that can provide data about the project and personnel. Two topologies describing the environment and this information integration process are now described. The first, shown in Figure 11.2, is an ideal situation where a central object oriented database (or single project model) is used (Anumba *et al.*, 2000). Here, all building elements are expressed as multi-attribute objects (including geometric data, cost, structural qualities, manufacturer, etc.) The adoption of this approach is a general aim for the industry and would allow users to generate discrete 2D and 3D drawings, documents, schedules, costs, etc. from the central object repository. Similarly, it would be possible to generate a 3D visualisation of the project from the same model with links to information objects in the model integral to it. This visualisation could then form the basis of a Telepresence Environment. The approach adopted in research projects such as OSCON could soon allow such visualisations to be produced with a minimum of human intervention (Aouad *et al.*, 1997).

It should be stated at this stage that a full 3D representation of the whole project containing all objects is not appropriate for use as a navigable 3D visualisation. The complexity and detail in such a model would be far to high for even the most powerful computers to provide a level of interaction that would be acceptable to users. Instead, it is necessary to provide a

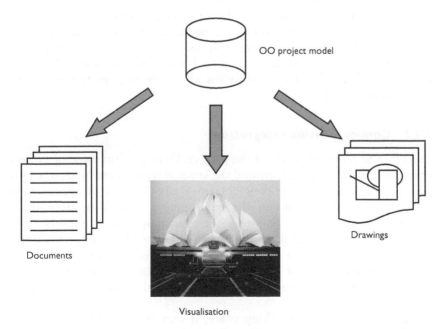

Figure 11.2 Ideal topology.

visualisation that is in effect an abstraction of the full model. Different views or layers can be used to encapsulate detailed information and conceptual representations of object groups can be used.

Were such complex 3D visualisations feasible today, a logical extension of the Telepresence Environment would be to enable it as an on-line design environment. Although this approach does sound superficially appealing, it would however be fraught with difficulties such as addressing the impact of any changes upon other parts of the model, notification of appropriate personnel and the visual representation of design data pending approval and design as-is. It is also questionable whether the scenario of multiple users meeting within a virtual environment to collaborate upon a design is appropriate. A more likely scenario would involve designers collaborating by sharing a design application possibly with the use of additional conferencing facilities. Design changes would then be represented in a model once the necessary checks and approvals had been made.

The second topology, shown in Figure 11.3, describes an approach that is more feasible using today's systems. Here a document management system is used to control discrete 2D and 3D drawings, documents, schedules, costs, etc. in the form of files. Obviously it is not possible to generate a visualisation from such data and so a separate 3D modelling process is required to produce it. This has the disadvantage of also being a discrete file (albeit with links to other data) and so would not remain up-to-date with the design. In order for such a visualisation to remain useful over time, it would need to be periodically updated or be sufficiently conceptual such that it remained a good representation of the project and its principal elements. Links to the document management system would need to be manually inserted into the 3D model. These would allow user to select objects in the 3D model and receive information about them from the document management system.

11.4.2   Communication integration

In addition to representations of the project, there are also those of project personnel. A user navigating around the space is represented by their avatar thus allowing others to see their current focus. The act of approaching and/or selecting individuals in the environment will allow appropriate integrated communication channels to be opened up. These channels could be textual, audio or visual depending upon different situations and requirements. This facilitates ad hoc project discussions. Individuals querying information systems about a common area of interest are able to interact with each other whilst using the relevant part of the model as a reference for discussions.

Once communication is taking place, it may be appropriate to represent this fact to other users. This allows them to request to join in the

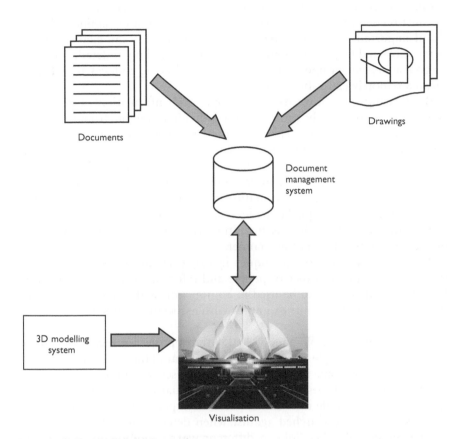

Figure 11.3 Currently feasible topology.

conversation if they so wish. Those currently communicating are able to accept or deny these requests.

11.4.3   Profiling and automatic navigation

Descriptions of functionality have, up until now, focused upon situations where the user is navigating around the environment by themselves. A parallel paradigm is that where the user's view of the environment is controlled. The reasoning behind this approach is that information can be conveyed to the user in a manner that promotes serendipity. These are the meetings that happen when two people meet in a corridor or other shared/public space. A discussion often ensues that at least one of the parties was not prepared for and generally the information gleaned from it was certainly not directly sought. Parallels can be drawn to situations where people are located 'on-site' or at the 'workface'. Chance encounters with product related objects or events can provoke thought or remind individuals of important issues. In general, chance encounters can enable a wider perspective and help to promote a better understanding of detail. They are, or course dependent upon the entities involved being co-located and those located remotely can suffer as a result.

By drawing users towards both people and information of relevance, chance encounters can be promoted for individuals remote from each other or the site. In fact, there is no reason why this should not be equally beneficial for those working in the same office or at the site itself since co-location does not necessarily infer that awareness already exists.

The important issue is how to determine the relevance of people to each other and the relevance of information to individuals. Similarly, how much information should people be made aware of? Too much will soon lead to information overload whereas too little will cancel out the need for the delivery of information in this manner.

There are a number of techniques in existence and under development that can allow the relevance of people and information to an individual to be determined. These can be used in conjunction with each other to build up a personal profile. Some examples with associated issues are:

- Recording the role of an individual within a project or specifying a name for their job. People often are quite particular about what they call their job and two people in the same job may describe it in differing ways.
- Recording a job description of an individual using free text or a list. This might include their skills, previous experience and current activities. It could be searched against when determining relevance. Again, people describe their jobs in different ways. Additionally, this sort of information can be difficult to collect and keep up-to-date.
- Specification of skills or interests against a pre-constructed list or ontology. This has the advantage of constraining people to a set of descriptors and makes subsequent searches against the list easier. The difficulty here can be the generation of the ontology – people will often want to add to or change it. Again, keeping the profile up-to-date can be an issue.
- Searching through a body of work to determine expertise or recurring themes. The work could be a series of documents or drawings authored by the individual or, as in the case of the MIT expert finder (Vivacqua, 1999), a body of computer code. Algorithms can be constructed to search through the work and generate a profile. This has the advantage that, since it is largely an automated process, it can more easily be kept up-to-date. However, it is complex to set up.
- Current focus. Systems can be employed to monitor the focus of the user. This may involve determining the Web page currently being

accessed, or monitoring the words being typed into the computer (Crabtree *et al.*, 1998). Again, this approach is automated and very powerful but also has inherent security and privacy issues.

* Recent focus. Similar to the above except that historical data is kept to maintain a wider view of interests and concerns.

These examples illustrate the possibilities but also identify the difficulties and research issues that are currently being addressed or need to be addressed in the future. For the purposes of this research it is necessary to select an approach or approaches that map well onto the problem domain and can effectively utilise the information systems being employed.

It is important to consider what kinds of information should be presented to users. It is envisaged that at any one time, there will be a number of entities that will be of relevance. These may be of the following nature:

* Other users with similar profiles;
* Other users looking at the same information;
* Objects relating to relevant data;
* Objects relating to recently generated events; and
* Other users who have either authored some documents of relevance or who have recently generated an event of relevance.

Events may be Requests For Information (RFIs), approvals, delivery notifications, etc. The relevance of each entity may change over time as might the 'need to know' about it, for example a drawing approval might be very relevant immediately after it occurs but become less relevant over time. Similarly the 'need to know' level of an approval will be largely reduced once it has been read or acknowledged. As deadlines for RFIs approach, their relevance and 'need to know' levels might increase. With people, colleagues who work very closely maybe highly relevant to each other but their 'need to know' about each other would probably not be high.

The result of these varying factors will mean that the information displayed to the user should change over time. Indeed, if the interface is to promote chance encounters, then changes of scene and visual stimulation through movement should be used to draw the attention of the user. Importantly though, it should also not be too repetitive and it should respond to user acknowledgement so that digested information is not repeatedly delivered (although some repetition of highly relevant scenes may be beneficial).

It seems that what is required is a constantly updating list of entities for each user. Each item on the list should have rationale as to why it is considered to be relevant and a measure of that relevance which changes over time. As each item is presented to the user, they should be given the opportunity of acting on it by initiating communication or retrieving information or merely acknowledging it. These actions will alter the relevance of the item or remove the item from the list.

11.4.4   Conceptual architecture

Figure 11.4 shows a conceptual architecture of the proposed Telepresence Environment (Anumba and Duke, 2000). The elements identified are considered essential in providing the functionality expressed in the previous section. Of the two approaches to information integration described in the previous section, the second was chosen for inclusion in the conceptual architecture. The reasoning behind this was that it relies on systems that are currently in use in the industry and could be implemented more readily that the first approach. The function of each element in the conceptual architecture is now explained as are the links to other elements.

11.4.4.1   3D model

As shown in Figure 11.3, the 3D model needs to be generated in a process separate from the detailed design (although it may already exist as a result of an early design process). Virtual Reality Modelling Language (VRML) is a standard commonly used for publishing 3D models over the Internet making it appropriate for use in this situation. Commercial products do

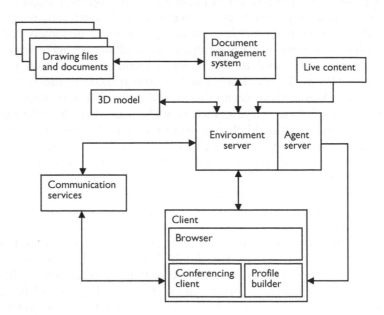

Figure 11.4 Conceptual architecture.

exist that convert existing 3D CAD files (such as DXF) into VRML. However, as stated above, design files are generally too complex to be used as visualisations and require skilled and time-consuming work to be carried out upon them.

An alternative is to use models built with the intention of being used for visualisation or to build one from first principles using a commercial package such as Concept CAD or 3D Studio VIZ. VRML allows URLs to be associated with objects. As well as being static, links to Web pages or models can be made dynamic. That is, the link contains descriptors for the item that can be used in a search of the document management system. For example, a beam object might be given a URL containing descriptors for its type – beam, its location – south wing and its work package – structural. When the object is selected the database can then be searched against the descriptors and a list of relevant information can be presented to the user. It is difficult to determine how effective such a process would be. It is heavily reliant upon the descriptions given to objects, upon the granularity of the objects and on relevant data being returned from a search (especially when much of the data being searched upon is graphical). In a scenario where an object-oriented modelling system is used, the process is more likely to be successful since a direct translation can be made between the visualisation object and the design object.

11.4.4.2   Environment server

The environment server is the central element in the system. It mediates the links between the different elements and serves the environment that the user sees at their browser. Users starting the client will logon to the environment via the server which will then manage the user session. The server is linked to the document management system and mediates the serving of documents to the user with authentication provided by the current session. In order to allow the control of the client view, the server obtains a list of relevant entities from the agent server, cycles through these and controls the client view appropriately.

11.4.4.3   Agent server

For each user, the agent server obtains a current profile from the client. It combines this with static profile data (such as job role, etc.) and uses it to determine relevant information from the document management system and relevant people. The details of these items are then made available to the environment server which uses them to control the user's view.

11.4.4.4   Profile builder

The profile builder monitors the PC usage of the user in order to determine current and recent focus and the activity of the PC. This is made available to the agent server which combines it with the static profile data that it stores. It may be beneficial to give the user the opportunity to add a drawing or element to their interest profile so that when changes or events occur in relation to it they are informed of them. This will be in addition to the formal process where the notification must be assured of getting through, for example, via e-mail.

11.4.4.5   Document management system

This element is necessary in order to handle all the formal processes that will remain in existence such as drawing change control, access permissions and sign-off. In order to be used in this architecture, it will need to Web-enabled, that is, all interaction should be possible via the use of a Web browser. User accesses of elements in the landscape will result in queries on the management system. The user will then be presented with a list of relevant drawings or documents that they can view in their Web browser or other viewing application. Authentication is handled via the user session.

11.4.4.6   Communication services

The environment server will control access to these services. It is aware of the relative position of users and the communications capabilities that they have (via the people database). Using these, it can set up appropriate communications links. The services will range from those with low cognitive load, such as text chat and e-mail for background conversation up to those with full cognitive load, such as audio and video conferencing for focused meetings. Interactions between individuals will be depicted within the landscape and other users will be free to join in conversations or request to join in meetings.

11.4.4.7   Client

The client is the main element on the client machine. It is the user's entry point to the system. It consists of a 3D viewer element to support interaction with the environment. In addition to this are elements capable of handling interactions with the server (such as authentication) and with data and communication tools on the client machine. It will also contain the profile builder.

11.5   Development and operation

11.5.1   System overview and development environment

The concepts proposed in the environment are by no means an incremental step from those that are employed in existing tools currently in use in the industry. For this reason, it is very difficult to assess whether the concepts are appropriate and that the proposed methods of implementation are correct. As a result of this the production of a concept demonstrator was identified as an appropriate course of action. The lack of available resources also meant that is was unlikely that a fully working system as described in Chapter 10 could be built and evaluated on a real project with a trial group of users. A concept demonstrator would allow the realisation of at least some of the proposed facilities and would enable an evaluation to be undertaken of the value of the proposed system.

Figure 11.5 shows the conceptual architecture introduced in the previous section. In this version, the various elements are colour coded depending upon their level of inclusion in the demonstrator. The concept demonstrator simulates those elements illustrated in blue. The simulation of the environment server is the main function of the demonstrator. It illustrates the log-on process, the representation of other users, the presentation of agent-provided information and the mediation of access to other people and information. The demonstrator also simulates the agent system. It would be difficult to otherwise illustrate their function within a stand-alone demonstrator. The agent software relies upon profiles built over a period of time as well as interaction with other users' profiles. The absence of other users in the demonstrator also means that there is no one to communicate with

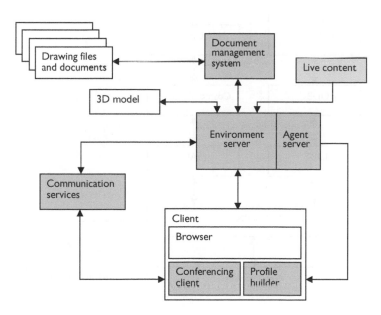

Figure 11.5 Architecture of demonstrator.

in order to demonstrate the communication services in a 'live' manner. For this reason, the communication technologies and others to communicate with are simulated by the environment. Finally a Web-based document management system is simulated in order to illustrate how it would be integrated into the environment. The items in white that is the browser, VRML model, and documents and drawings are principally included 'as-is' in the demonstrator. The item in grey (i.e. Live Content) has not been included in the demonstrator due to the lack of time.

Figure 11.6 shows the technologies that were employed in producing the demonstrator. These are described in turn.

Access database   Microsoft Access is the database component of the Microsoft Office suite of software. It allows the creation of multi-table relational databases. This means that complex sets of data and the relationships between them can be modelled and stored. It is more than adequate to satisfy the performance requirements of the prototype, is readily available and was chosen for these reasons.

The Open Database Connectivity (ODBC)   is used to provide a link between a proprietary database and the generic Structured Query Language although in this case a further technology is required to enable that interface – JDBC (explained in the next section).

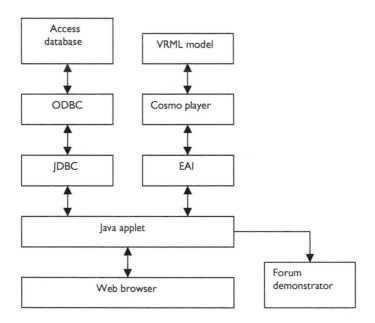

Figure 11.6 Technologies architecture.

Java Database Connectivity (JDBC)   is the technology that provides this extra link. This set of libraries is required as an addition to the Java language which has no in-built SQL or database support.

Java Applet   The Java applet is the core of the demonstrator. It provides the interface between the user, the database and the VRML model. It was decided to implement the demonstrator using Java for a number of reasons. First Java has a well documented interface with VRML – vital for manipulation of the model itself and the user's view of it. Second since Java is interpreted rather than compiled code, it can in theory be written once to run on any platform. Thus any device with a Java Virtual Machine can interpret the Java byte code and run the application. Finally, a Java applet is downloaded and run within the context of a Web browser. This has the advantages of users not having to install an application or to re-install an application when changes or bug-fixes are made. The disadvantages of using a Java applet are that they run more slowly than a compiled application and that they require downloading before they can be used.

VRML model   Virtual Reality Modelling Language (VRML) is the 3D modelling language for the Web. This has previously been described.

Cosmo player   Cosmo player is one of a small number of VRML 'plug-ins'. Basically it provides the Web browser with the capability to display VRML models. It also provides its own user interface that allows manipulation of and navigation around a 3D environment.

Extended Authoring Interface (EAI)   EAI is the link between Java and VRML. It allows a VRML model to be controlled from within a Java program and also allows events generated by VRML to be passed back into a Java applet or application.

Web browser   The Web browser is the environment in which the demonstrator is run. Most Web browsers contain a Java Virtual Machine which allows them to interpret and run Java byte-code. It also provides the so called 'sand box' in which the applet runs separated from the local file system thus providing a level of protection for it from the downloaded code.

Forum demonstrator   The Forum demonstrator is the stand-alone demonstration of an Audiographic Conferencing system. It is employed here as an illustration of how communications facilities can be integrated into the environment. The demonstrator is an installed application and does not run in the Web browser.

11.5.2   System development

The first phase of the system development involved determining what exactly the demonstrator should do. This was achieved by 'storyboarding' a typical user session. It was recognised that this storyboard would also form the basis for the demonstration of a typical user session once the prototype was complete. As well as this, it also formed the basis of the

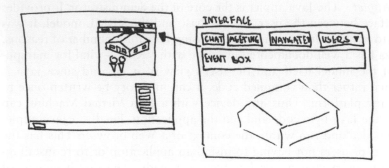

Figure 11.7 Example storyboard scene.

development plan with each feature illustrated by the storyboard being developed and tested in turn. An example of the hand-drawn images from the storyboard has been reproduced here with the use of a scanner and can be seen in Figure 11.7. The storyboard shows a user's desktop with various windows open on them.

11.5.2.1   The 3D model

The VRML model used was supplied by W. S. Atkins. The model was produced using 3D Studio Max and output into VRML using Max's VRML97 Exporter. The building modelled is a planned school extension. It can be seen in Figure 11.8.

It was explained earlier that the environment relies on the various work packages of a project to cluster people and information within it. The following work packages were identified:

• Foundations
• North Wing
• South Wing
• Roof
• Doors and windows.

For each of the work packages, a location and orientation within the model were defined. These were situated near and facing the features that the work packages are related to. As such, they became known as feature

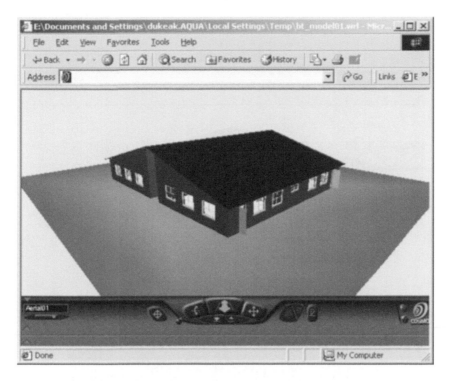

Figure 11.8 Building model.

locations. Obviously, the feature location for the doors and windows work package could have been situated in a number of locations since there are several doors and windows. There may be many more such work packages in a real project. An arbitrary decision was made as to where this feature location should be situated. As long as the placement of such locations is consistent, this is not considered to be a problem since users should quickly become used to what is situated where.

11.5.2.2   Avatars and the user's view

For the purposes of the demonstrator very simple avatars are used. They are just a sphere and cone combination of varying colours. However it is debatable whether the avatars in a fully working version would need to be significantly more complex. Figure 11.9 shows an avatar in the environment. The user has their mouse pointer over the avatar (although this is not shown in the screenshot), which results in the name of the person whose avatar is being displayed in the status area of the environment. A touch

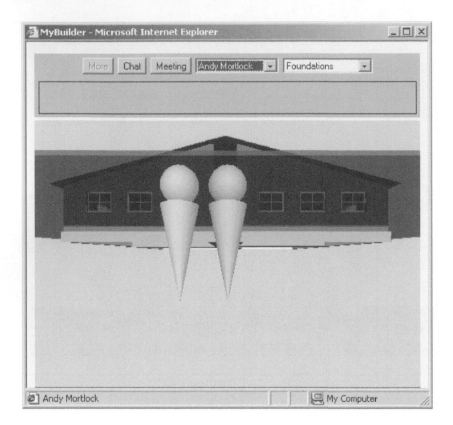

Figure 11.9 Mouse over an avatar.

sensor was also included with each avatar. When activated, the homepage of the user in question is displayed. This was carried out by querying the 'people' database for the user id of the clicked avatar to return the URL of their homepage.

11.5.2.3   Moving around the virtual world

There are two ways in which a user's position in the environment is changed. The first is where the user is moved by the agent as a result of an event. The second is by user intervention that is where the user elects to move to a different location or elects to move to the same location as another user. This is achieved by the user selecting a zone or other user from one of the pull down lists. The user list is kept up-to-date with all of the other users who are currently logged on. When another user is selected, the 'users' table of the database is interrogated to find out at which location they currently are. The user is then moved to join the selected user.

11.5.2.4 Presenting events

One of the major features of the environment is the subtle presentation of information to the user via a movement to a feature and a short statement of rationale. This information is generated by the agent. In the demonstrator the action of the agent is simulated.

When an event is presented in this manner the 'More' button on the interface becomes enabled. Thus if the user decides that the event is of interest they can press this button to get more information about the event. A separate window is opened if this button is pressed. This is shown in Figure 11.10.

The window allows further information and options that the confines of the environment window do not allow to be presented to the user. Further information about the event could be presented here, such as why the agent believes that it is relevant to the user. Additional choices are available via the buttons in the notification window. These are 'Open Document', 'Open Author's Homepage', 'Navigate to Author', 'Delete This Note' and 'Close'. The first two options will open up a separate browser window containing the document or the homepage. The 'Navigate to Author' button is only enabled if the author is currently logged in to the environment. This is included should the user wish to contact the author via the environment. Once the user has read and dealt with the event, they can delete it. This removes the link to the event from the 'notification' table mapping the event to the user. Following this, the event will not be presented to the user again in future.

11.5.2.5 Simulating other users

Of equal importance to the presentation of events, is the interaction that is enabled by the fact that the environment is a multi-user one. This is simulated in the demonstrator with the use of dummy users that move around

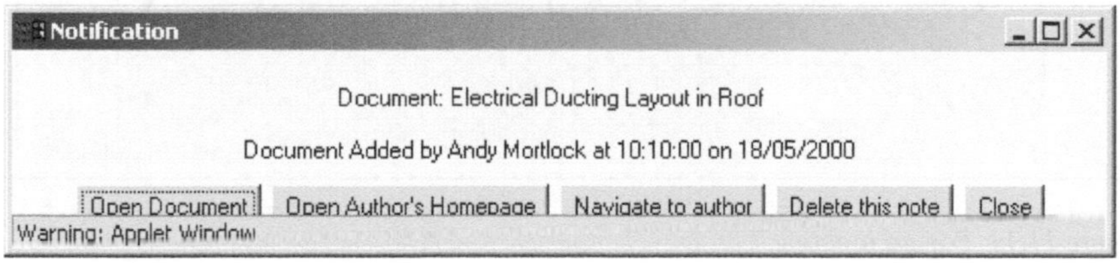

Figure 11.10 Notification window.

the environment between feature locations in the same way as the user's avatar. Each avatar has two features that users can make use of. By moving their mouse pointer over the avatar a user can find out whom the avatar is representing. Their full name is displayed on the status area of the interface. By clicking on the avatar the Web page of the user is opened in a separate browser window.

11.5.2.6   Integrated communication

The third major feature of the environment is the facility for users to initiate instantaneous communication with each other via the environment. In the demonstrator this is illustrated with two forms of communication. The first is text chat. This is intended to support informal ad hoc communication. The second is the Forum Meeting Space (Figure 11.11). This is intended to support more formal and/or richer forms of communication. The user interface includes buttons that allow both methods of communication to be initiated. These are enabled whenever there is another person in the same feature location as the user. Pressing one of the buttons brings up a selector box with a list of people in the current location. The user can

Figure 11.11 Forum meeting.

then select which of these people they with to invite to a meeting. Following this an invite is sent out to those selected, which they can choose to accept or reject. In the demonstrator the dummy users automatically accept the requests. In the case of the text chat a chat box is then opened up which the users can add text to. With the 'Meeting Space', the stand-alone demonstrator is started. This is a separate application that illustrates the facilities in that system (such as symbolic acting, application sharing, audio conferencing, etc.).

11.5.2.7   The Document management system

A simple document management system was added to the demonstrator. This illustrates two features of the environment. The first is having the documents integrated into the environment. When the user clicks on a feature in the VRML building model a database query is fired off. The 'documents' table is queried for all of the documents associated with the work package that the feature represents. Details of these documents are listed in a window representing the document management system. The user can select one of the documents in the list and open it. This launches a separate browser window with the URL of the document as shown in Figure 11.12.

A number of drawings related to the school building were supplied by W. S. Atkins and these were added to the 'documents' database along with 'Microsoft Word' and 'Excel' files representing other types of document. When one of these documents is opened, an appropriate browser component object handles its display in the browser. This was a .dxf viewer in the case of drawings or the appropriate Word or Excel control.

The second feature illustrated is an example of how the agent might recognise general user activity and navigate their avatar appropriately in the environment. This simulates normal use of the system. When the user carries out a search on a particular work package or opens a document related to a particular work package this is recognised and the user's avatar is navigated to the feature associated with that work package. This could be extended to navigate to a user when a search is made by name. This is a very simple demonstration but is intended to promote the awareness of the link between user activity and the environment.

11.6   Evaluation

An evaluation exercise was a carried out with the aim of determining how appropriate the application of a 'Collaborative Virtual Environment' to support distributed teams in CLDC projects is.

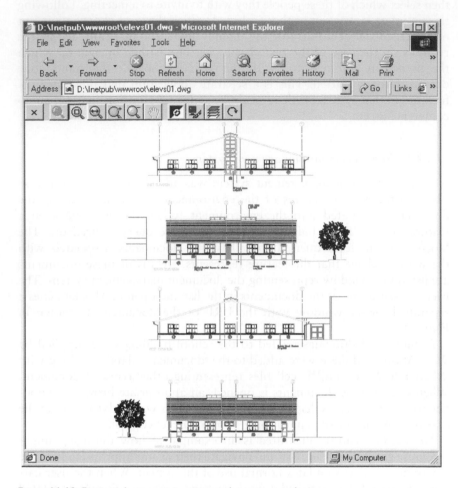

Figure 11.12 Project document in separate browser window.

Specifically, the objectives for the evaluation were:

- To assess how well the environment and the features within it, support the notion of co-location within a virtual space that is Telepresence;
- To determine how well the environment encapsulates and represents the project itself;
- To assess whether the various elements of functionality associated with the delivery of information about people and project data are both appropriate and suitably well implemented;
- To assess whether the communication tools in the environment are sufficient and suitably integrated;

- To determine how well the environment allows users to improve their awareness of activities on the project and whether this improves the ability of the user to fulfil their role;
- To assess whether the adoption of the environment could improve the prospects of a project achieving the goals of CE;
- To determine whether industry practitioners would make use of the environment and more specifically whether any groups or professions would be more or less likely to do so.

11.6.1   Evaluation process

Following the decision to produce a demonstrator for the 'Telepresence Environment' it was necessary to evaluate it in line with the objectives outlined earlier. This was carried out by presenting the demonstrator to a number of industry practitioners and obtaining their feedback via a questionnaire. All those who were invited to participate had practical experience within the industry as well as experience of 'Information Technology'. In all, the opinions of twelve people were obtained.

In order to ensure that the respondents were familiar with the concepts involved, they were first shown a presentation. This explained the concepts of Telepresence and CE. It also introduced the Telepresence Environment and identified its broad objectives.

The demonstration immediately followed the slide presentation. This was carried out with the use of a laptop computer connected to a projection system in front of all the participants. The participants were invited to interject at any point during the demonstration to ask questions of clarification. The demonstration itself took the form of a sample usage session. This covered the main points of user interaction in an attempt to illustrate its features. After the demonstration there was a further opportunity for questions of clarification. Following this, the participants were requested to complete the questionnaire.

11.6.2   Evaluation analysis

The evaluation process resulted in a body of quantitative and qualitative results. The quantitative results were in general positive and give a good indication that the prototype is considered to be appropriate to the industry and that the application of the concepts it illustrates would be beneficial. The assessment process was successful in that it provided an indication of how appropriate the prototype was, as well as important issues that need addressing and areas in which it could be improved. However, from the results themselves it is clear that the assessment process alone is insufficient to determine the full suitability of the prototype and the concepts it illustrates.

11.6.3   Evaluation results

This section distils the questionnaire responses and presents the potential benefits and limitations of the prototype environment based upon the evaluation. The benefits are:

- The environment and the features within it, support Telepresence;
- The methods used to deliver information about people and project data are both appropriate and suitably well implemented;
- The environment provides communication tools that are useful and well integrated;
- The user is supported in fulfilling their project role and has an improved awareness of project activities;
- The adoption of the environment would improve the prospects of a project achieving some of the goals of CE;
- The environment would be widely used by design and management teams.

The limitations of the demonstrator are:

- The environment does not contain realistic avatars;
- The project could be represented better by adding further features to the environment;
- The environment could increase the problem of information overload rather than reducing it;
- The prototype does not provide tangible evidence that the adoption of the environment would improve the prospects of a project achieving all of the goals of CE;
- The low use of IT and the lack of skills (particularly in site roles) are potential barriers to adoption.

The potential benefits of the environment as given by the respondents provide a positive assessment for each of the objectives laid out earlier. There were also some very valid limitations expressed which are now discussed. The first of these were that the avatars and features in the environment could be improved and extended. These are concerned with technical aspects and the suggested improvements could be readily integrated into future implementations should they prove to be beneficial. However, as stated above, improved avatars would not necessarily improve the quality of the environment as a whole.

Information overload and the lack of IT skills are both industry issues that could affect the success of the environment. The environment is attempting to reduce the problem of information overload. Great care should be taken to ensure that is not increased and this should be monitored in any future evaluation processes involving users. However, the nature of the environment is that users should be able to make as much or as little use of it as they see fit. The latter issues of low IT skills and usage are well recognised and represent a major problem for the adoption of technology in general. However, the signs are that the necessary improvement in infrastructure is beginning to take place. This is also reported to be the case with the usage of some key technologies and importantly the attitudes and acceptance levels of members of the industry.

The vital way to improve the prospects of the adoption of a technology is of course to prove its benefits in relation to cost (and time). As a result, the limitation of the environment in failing to provide evidence of an improved project is one that definitely needs to be addressed. One way that this could be addressed is to trial the environment in a real project situation. This would give project personnel the opportunity to use the environment over an extended period of time, get used to its functionality and form an opinion on whether the stated benefits are actually achieved. This is certainly not an easy task as there are many difficulties involved with carrying out a trial in a working environment. The difficulties are magnified when, as with the Telepresence Environment, the trial system must integrate with existing systems. The remaining limitation that was expressed was that the environment could add to the problem of information overload. Again, only a trial situation would allow the users to determine whether the problem was improved or worsened by the Environment; however, the issue should be considered at the forefront of further development activities.

11.7 Conclusions

The following conclusions can be drawn from the research:

* The Telepresence Environment is a unique and innovative approach to enhancing collaborative communications on construction projects. It is a medium that allows non-collocated construction personnel to collaborate at a level approaching that of co-located colleagues. Chance encounters are seen as vital to the success of an organisation. This unplanned form of communication occurs frequently when colleagues are co-located or are located at the 'workface' (i.e. the construction site in this case). The Telepresence Environment allows this form of communication to occur in the disparate organisations that are common in the construction industry. It promotes to the user an awareness of other people or project information that it deems to be of interest to them. The user is then able to act upon this 'chance encounter' via the environment. As well as acting as an integrator for people, the environment also integrates communication and project information services into a common user-interface. This allows the user who wishes

to act upon a 'chance encounter' to easily access the appropriate project document or drawing or to speak to and collaborate with the appropriate people. The improved collaboration and communication that is enabled by the Telepresence Environment will allow the decisions made by project personnel to be more informed. This will in turn lead to an improved product with fewer mistakes made.

- Telepresence technology is an effective way of facilitating communication within a CLDC environment. The adoption of CE within the construction industry will increase the level of communication that is required between the parties involved. The nature of the industry is such that the parties involved in projects will generally not be co-located throughout the design and construction phases. As a result an increased reliance is placed upon telecommunications technology. Telepresence is one technology that is particularly suited to adoption in this environment. It provides users with a shared 3D environment that allows virtual co-location and the ability to interact with virtual objects within it. Construction is inherently three-dimensional in that it is concerned with the production of complex 3D facilities. As such, a technology that can represent facilities in the same manner is appropriate. The importance of chance encounters to an organisation was identified above. These are of particular importance in a CLDC environment where greater co-ordination is proposed along with early problem discovery and early decision making. Telepresence allows these encounters to be enabled in an unobtrusive way with the use of movement around a 3D space and toward recognisable objects within it.

- A telecommunications infrastructure for CE in construction provides an effective means to deliver the necessary improvements in communication. The importance of communication to the success of CE has already been established. Telepresence is one communications technology that is appropriate but there are several others that should also be applied such as video, audio and data conferencing, virtual and augmented reality and project models. An appropriate telecommunications infrastructure will enable all of these technologies to be deployed in an integrated way. From the user's perspective this involves an integration of the services at the user-interface, allowing them to switch between appropriate communication channels as they see fit. The emergence of Web-enabled services has provided the means for this to be delivered. The infrastructure also provides a level of 'back-end' integration. This allows project information systems to inter-work with communications services providing capabilities such as universal directories and authentication of users and access to necessary project information from within a communication service.

11.7.1   Further development

In order to evolve from the concept demonstrator into a fully working system that can eventually be trialled in a real project setting, there are a number of steps that must first be carried out. The first of these should be to carry out further software development work to fulfil the requirements of the conceptual architecture. The four components of the architecture that require development are:

- *Environment server*: This needs to be developed into a component that can store the state of the environment and can send and receive event messages to and from the various clients that are connected to it. It will also need to act upon events received from the agent server. An existing 'Collaborative Virtual Environment' known as the the 'Forum Contact Space' contains a server component that carries out most of the required functions and as such would be appropriate for use in this situation. It would need to be adapted to serve the virtual world developed for the prototype and to deliver the necessary project-based data.
- *Client*: This needs to be developed to send and receive event messages to and from the server. As with the server, the 'Forum Contact Space' has a client component that could be adapted for use in this situation. Again, the set of events that it handles would need to be extended, as would the methods for the presentation of data.
- *Agent system*: The 'Forum Contact Space Agent' system had already been suggested as a candidate for the required Agent Server and Profile Builder. The case for its use is further strengthened if the other Contact Space components are employed since it can easily integrate with them. It too would require adapting in order to cope with the meta-data obtained from documents accessed in the 'Document Management System'.
- *Communication services*: A simple text chat system like that illustrated in the prototype is already provided in the Contact Space. Other services such as 'Audiographic conferencing', provided by the 'Forum Meeting Space' can be readily integrated, particularly as this technology is moving towards a Web-based delivery mechanism. Requests to set-up a meeting would be handled by the Environment Server which would communicate with a Conferencing server via an API.

Once a fully working system is available, further evaluation processes can then be carried out upon it. A short 'hands-on' evaluation session with industry practitioners followed by an interview would give limited information about the broad benefits of the environment due to the lack of co-users and real project information. However, this style of session could still provide some benefit. It would enable information about for example, the quality of the user interface, the communication services and the style of information delivery to be acquired. These factors are important and could improve the chances of eventual adoption by industry practitioners.

There are specific research areas that only a trial in a real project setting would be able to provide sufficient evaluation of. These are identified below with a summary of what further research is required:

- The Environment currently uses a single model to represent the project. This provides a fairly high level view with sub-classifications of it based around major features for example South wing. The trial should attempt to evaluate whether this method is sufficient or whether improvements could be made that could inject more meaning from the point of view of the various disciplines (e.g. architectural, structural). It may be appropriate to allow members of the different disciplines to select their preferred view of the model.
- The generation of user profiles within the Environment is carried out by a modification of the agent system used in the Forum Contact Space. The modification consists of allowing the collection and processing of meta-information from the documents and drawings accessed by the user. The trial and evaluation should consider whether this method of maintaining user profiles is appropriate. This might involve allowing those involved in the trial to express an opinion on whether events that they are made aware of by the agent system are considered by them to be relevant.
- The links to underlying data in the model are currently based on classifications of work and trade package from the document management system. This method relies upon these classifications being at a sufficient level of granularity such that objects in the model can be assigned to them. This approach is based-upon the 'currently feasible topology' shown in Figure 11.3. The full-scale trial process would allow the approach to be evaluated, again using feedback from the users concerning the quality of links between the visualisation and design information. Developments such as those in the OSCON project (Aouad *et al.*, 1997) may make it possible to automatically generate these links. It may be appropriate to explore this approach in future research on the Telepresence Environment.

11.7.2   Developments in Telepresence for construction

As well as the further developments that are applicable to the prototype, there are also a number of more general areas in the field of Telepresence that require attention. Some of these are now briefly discussed.

The deployment of a network infrastructure that is sufficient to deliver, to projects, the services described here should be carried out. The network requirements of these services are an appropriate topic for further research. Many players in the industry are making the necessary investment in this area. The deployment by the telecommunications industry of technologies such as Asymmetric Digital Subscriber Line (ADSL) and cable modems should allow rapid strides to be made in enabling high-speed project Extranets to be set up.

Investment should also be made in the services that these networks will enable. Unfortunately the up-front costs that are associated with ICT investment can be prohibitive, particularly to the smaller players in the construction industry or where projects are of a duration that is deemed too short to make the investment cost-effective. The Application Service Provider (ASP) market is one that is experiencing high growth at present. This market basically involves an organisation renting services from a third party solution provider and thus massively reducing the up-front costs. These services might consist of managed computer or network hardware or network-based supported software solutions. This market is highly applicable to the construction industry because of the short-duration projects and small companies that are often involved in them. The delivery of Telepresence and related technologies via this scenario is one that could be researched further.

The Telepresence Environment described here is designed for desk-based users with their own computer. This scenario is only applicable to a small proportion of the personnel on a typical project. Many people will spend a large amount of their time away from their desk, either travelling to meetings or on-site. If the environment is to support people whilst they are away from their desks then an additional form of information delivery is required. Mobile Internet services are beginning to emerge. At present, Wireless Application Protocol (WAP), services are available that provide text-based Internet services on a mobile phone. In the near future, so called Third Generation (3G) mobile networks will greatly increase the bandwidth available to mobile users. Much richer services will become available with hand-held computers or Personal Digital Assistants (PDA) becoming the device the mobile user accesses them with.

The delivery of project-related information to the mobile user would be a worthwhile topic for future research. This could begin with providing text-based information over WAP and extend to include richer services such as data or video in preparation for 3G mobile. These developments also increase the potential of technologies such as 'Augmented Reality'. One aspect of AR might be to provide project information to a mobile user from a network database. This could be delivered to the user over a high bandwidth mobile network removing the need for a wireless LAN to be in place to serve them.

The research project on which this chapter is based has largely focused on the technical aspects of the Telepresence Environment. An additional research area that should be addressed is that of the human factors associated with its deployment. One of the main aims of the project is to change the way people in the industry work, moving away from the insular nature of traditional processes such as 'over the wall' towards a more open, collaborative approach. This shift requires a change in culture to both the organisations and the people working in them. Such changes can rarely occur without difficulties. A Human Factors study should attempt to identify issues related to these changes as well as other issues such as the usability of the technology.

In the light of the research presented in this chapter and the developments suggested in this section it is recommended that the construction industry make wider use of collaborative communications technologies and in particular, Telepresence. A number of technologies have been identified that could and should be readily applied to construction projects such as the People and Information Finder, Conference Call Presence and Web-based Document Management Systems. The industry should continue its increased investment in technologies such as these and in the networks and infrastructure that are required to support them. The improved infrastructure will also allow more advanced technologies to be implemented. One such technology is, of course, Telepresence and its potential benefits have been clearly stated in this chapter. Some of these benefits are as yet unproved and as a result the industry should carry out trials and research in order to make informed decisions about the development and eventual adoption of Telepresence.

11.8   Note

1 VisionDome is a registered trademark of Alternate Realities Corporation.

11.9   References

Anumba, C. J. and Duke, A. K. (1997), 'Telepresence in Virtual Project Team Communications', *Proceedings of the ECCE Symposium on Computers in the Practice of Building and Civil Engineering*, Lahti, Finland, 3–5 September, pp. 80–84.

Anumba, C. J. and Duke, A. K. (2000), 'Telepresence in Concurrent Lifecycle Design and Construction', *Artificial Intelligence in Engineering*, Special Issue on Collaborative and Concurrent Engineering, Vol. 14, No. 3, pp. 221–232.

Anumba, C. J. and Evbuomwan, N. F. O. (1999), 'A Taxonomy for Communication Facets in Concurrent Life-Cycle Design and Construction', *Computer-Aided Civil and Infrastructure Engineering*, Special Issue on 'Collaborative and Concurrent Engineering', Vol. 14, pp. 37–44.

Anumba, C. J., Bouchlaghem, N. M., White, J. and Duke, A. K. (2000), 'Perspectives on an Integrated Construction Project Model', *International Journal of Co-operative Information Systems*, Vol. 9, No.3, pp. 283–313.

Aouad, G., Child, T., Marir, F. and Brandon, P. (1997), 'Developing a Virtual Reality Interface for an Integrated Project Database Environment', *Proceedings of the IEEE Conference on Information Visualisation*, London, UK, 27–29 August, pp. 192–197.

Cochrane, P., Heatley, D. J. T. and Cameron, K. H. (1993), 'Telepresence – Visual Telecommunications into the Next Century', *Proc. 4th IEE Conference on Telecommunications*, Manchester, April 1993, pp. 175–180.

Crabtree, B., Soltysiak, S., and Thint, M. (1998), 'Adaptive Personal Agents', *Personal Technology Journal*, Special Issue on Personal Information Agents, Vol. 2, No. 3, December, pp. 141–151.

Duke, A. K., Bowskill, J. and Anumba, C. J. (1998): 'Telepresence-Based Support for Concurrent Engineering in Construction', *Computing in Civil Engineering*, Proceedings of the ASCE International Computing Congress, Boston, October, pp. 549–560.

Madigan, D. (1993), 'BICC's Experience of Multimedia Communications: Issues in CSCW', *Proceedings IEE Colloquium on Multimedia and Professional Applications*, London, February.

Morris, C. (Ed.) (1992), *Academic Press Dictionary of Science and Technology*, Academic Press, London.

Rogers, A. S. (1994): 'Virtuosi – Virtual Reality Support for Groupworking', *BT Technology Journal*, Vol. 13, No. 3, pp. 81–89.

Traill, D. M., Bowskill, J. M. and Lawrence, P. J. (1997), 'Interactive Collaborative Media Environments', *BT Technology Journal*, Vol. 15, No. 4, pp. 130–139.

Vivacqua, A. S. (1999), 'Agents For Expertise Location', *Proceedings 1999 AAAI Spring Symposium on Intelligent Agents in Cyberspace*, Technical Report SS-99-03, Stanford, CA, USA, March, pp. 9–13.

Walker, G. R. and Sheppard, P. J. (1997), 'Telepresence – The Future of Telephony', *BT Technology Journal*, Vol. 15, No. 4, pp. 11–18.

Chapter 12

Support for users within a Concurrent Engineering environment

Raimer J. Scherer and Žiga Turk

12.1 Introduction

Working together is one of the most elementary human activities. It is possible in an environment that enables:

- sharing: all involved share the object of work,
- awareness: the involved workers are aware of what others are doing. They have a rough idea of who is doing what and how busy she is,
- coordination and control: the work is commanded; the level of control may be inverse to the level of awareness; higher levels of awareness require less control.

The prerequisite for all of the earlier is the possibility to communicate. If the object of work is information (such as a design or a plan) the communication is not restricted to person-person communication (for awareness and coordination) but also as a means to share and exchange the object of work. Finally, to get any work done, the people involved need access to some tools. The features earlier are needed in all working environments regardless of the technological complexity. For example for a team of chefs preparing a meal or for a master architect and his apprentices developing a plan of a cathedral in a traditional design studio. However, in those cases we would not consider them 'users' but cooks, architects, workers. They are not 'users' because there is no technology supported environment to be the users of. There is no need for such an environment, because the workers are co-located in space-time.

They become users if there is an information system or technology providing the environment. The main role of this environment is to re-create the space-time co-location – to enable the sharing, awareness and coordination event if they work at different locations and at different times.

Such an environment was crated and demonstrated in the 5th Framework Information Society project (2001–2003) called ISTforCE. ISTforCE was a 27-month EU 5th framework IST project, running from February 2000 through to April 2002 with an overall budget of about 4 million Euro. The partners came from Germany, France, Italy, Spain, UK, The Czech Republic and Slovenia.

12.1.1   Context and related work

In the second half of the 1990s the Internet has been intensively explored as a platform, which could be used to exchange information between architects, engineers, construction managers and the construction companies on the building site (Turk *et al.*, 2000b; Ouzounis, 2001; Weisberg, 2001). However, the Internet is still typically used to support only the non-core, non value-adding activities in the construction value chain. It is increasingly used as a communication platform (email) and a source of information (Web pages), but it has not yet been used as a place where the actual engineering work is carried out. Indeed, several companies have started offering Internet based project support on a rental basis. On Web sites, such as Bricsnet, Citadon and several others virtual companies can rent shared project space, with functionalities for publishing and retrieving design files, establishing security and access rights, versioning and configuration management, redlining, safe communication channels, mailing lists, notifications etc. (Bricsnet, 2001; Buzzsaw, 2001; Citadon, 2001; Conject, 2001; Weisberg, 2001). However, in spite of many advanced features, this is still only infrastructure – practically no tools are available that can actually get the engineering or architectural work done – work that contributes to the evolution of the design or process plan. This is due to the fact that these sites are project centred – they allow creating the support for one or several projects, but most engineers in construction practice work on several projects at the same time. Also, they only allow for file/document level information exchange, but can hardly manage more structured project information.

12.1.2   Objectives

The major project goal is to provide an open, human-centred Web-based collaboration environment maintained by a service provider, which supports Concurrent Engineering (CE) while working on multiple projects simultaneously and offers easy access to specialised engineering services distributed over the Web on rental basis. Normally, engineering applications are bought and then installed and used locally, but in the last years there is a growing interest, especially by small, highly specialised vendors, to offer such applications on rental or 'pay per use' basis. ISTforCE will provide them appropriate market access. However, in spite of certain marketing advantages of this new business model, there are also several problems that need to be dealt with. On the side of the service providers such problems include: how can the offered services reach the end users, how can they be paid for, and how should they be integrated into larger collaboration platforms to provide added value to the customers. On the side of the project management, appropriate control of service usage, monitoring of the eligible costs and certain legal aspects are issues to be cared about.

The main innovation of ISTforCE is in the *human-centred approach* enabling the integration of multiple applications and services for multiple users and on multi-project basis – to support the work of each user across projects and team boundaries and to establish a common platform where providers of software (often specialised SMEs) and end users (engineers, architects, technicians, project managers, etc.) can meet. The features of the platform, extending existing Web-based solutions, are as follows:

- *It is open*, so that any service or tool, Web or workstation based, could be integrated into it. Current collaboration portals offer only a small, fixed set of such functionality.
- *It is customisable to individuals*, so that several AEC professionals with different personal and professional requirements can use it in their own fashion. In contrast, levels of customisation of current solutions are only 'screen deep'.
- *It is customisable to products*. Each engineering of a construction product may require different IT services and tool infrastructure because construction products are unique.
- *It is customisable to projects*. Each construction project may require a different IT logistics infrastructure because projects are unique as well.
- *It is scalable*. Often, companies with different IT infrastructures take part in an AEC project. The platform is usable both on modest as well as advanced state-of-the-art equipment and networking speeds.
- *It is available*. By setting up an infrastructure for renting engineering and infrastructure software, thin clients as well will have access to advanced software and make it available to every size of company.
- *It is attractive to information providers*, easy-to-register and easy-to-plug-in interfaces will motivate them to provide their services into the platform, hence enhance the platform power and thus make their services available to a broad range of projects.
- *It is rentable*. Only a few companies can afford buying an expensive program just in case they might need it in the future or just once. Also, setting up IT is not the core business of design companies. They should be able to rent required infrastructure and be allowed to tailor it to their own specific needs.

With these features two important particular needs of the user in the construction industry have been met:

1   Multi-project work. In construction, engineers typically participate in several virtual enterprises in parallel, working concurrently on several

projects at the same time. This aspect of construction IT is in strong contrast to other industries and requires appropriate new solutions for collaboration.

2   Individuality. Due to their different roles, expertise and personal experience engineers have different individual preferences and capabilities of how to work and be creative. They need services that can seamlessly support their individual working in flexible and adaptable fashion.

12.2   Architecture of the platform

This section introduces the architecture of the platform developed in ISTforCE to support the users in CE work.

12.2.1   Users

Besides the typical end user, namely an engineer or architect, also service and tool providers will benefit from the Concurrent Engineering Service Platform (CESP) by easily offering their services and tools via CESP. Chief information officer can easily configure services, tools and servers for particular users and projects, that is to extend and modify user specific CESP for newly added projects. Project managers can easily get information about projects as a whole and intervene easily in project-centred work-flows. Hence, the platform is a place where four main groups of people with different goals will come together (see Figure 12.1):

- *Engineers and architects.* The platform not only supports the collaboration functions (sharing, awareness and coordination) but also serves as a communication backbone and provides access to the design and planning tools.
- *Project managers.* They need to manage projects, monitor architects and engineers, assign and supervise tasks, as well as tools and services. They mainly use the collaboration features, particularly the coordination aspects of it.
- *Chief Information Officers (CIO).* They want an easy way to set up and maintain the ICT infrastructure for a project. Again personalised access interfaces are appreciated.
- *Service and tool providers.* They need channels through which they can sell their products. They know their trade, but may not be experienced with Internet tools. They expect an infrastructure they can rent and which provides flexible and easy-to-understand APIs[1] so that they can concentrate on their core business.

The users of the platform can access the services in four ways discussed in the next section.

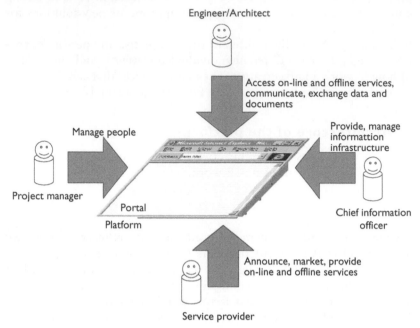

Figure 12.1 The four types of users of the platform.

12.2.2 Layered perspective

The typical architecture of current collaboration platforms for engineers is presented on high conceptual level on Figure 12.2. Through a Web browser on a user's desktop there is an interface to the rented project space. This project space typically includes file level information exchange, messaging and scheduling. A fixed set of services is available by the provider of the platform. TCP/IP is used to move data around.

ISTforCE extends this architecture in two ways. ISTforCE provides a well-defined, customisable architecture for the user desktop, suitable for the needs of the actors in the construction industry. The components of this 'personal concurrent engineering service platform' (pCESP see later) are flexibly adaptable to the needs of the end user. There the user can access the services through four different interfaces.

The second ISTforCE's extension is that ISTforCE provides three distinct, clearly defined layers for the collaboration platform (see Figure 12.3), that is the external (rented) engineering services, the specific extended infra-structure services and the core information services, which are open and accessible to the four different types of user groups instead of the closed system of the dotcom companies.

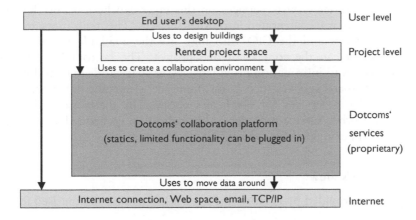

Figure 12.2 Typical current architecture of services for engineers on the Web.

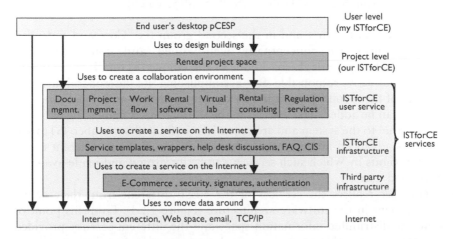

Figure 12.3 ISTforCE architecture. Three kinds of services are made available to be plugged in on three distinct layers.

The topmost of these three ISTforCE layers includes most services available to the end user, such as scheduling, document management, file and object based data exchange. This layer will be typically used by the managers and the engineers and architects for their actual project tasks. The middle layer includes services that assist either in the creation or help during the operation of the top-layer services. They are used both by the end users as well as by the providers of the services. An example of such a service is the help desk system or templates for document management,

FAQ, conferencing applications and the Core Information Service. They can augment any other service. Services on this layer can be used by the end users only if they are integrated into other services. The main users are the service providers and the CIO who may use the available templates to create a service in the context of one project. The bottom layer offers generic services to information providers (server space and email addresses, but also advanced services such as e-Commerce, electronic payment, etc.), as well as basic services for the layers above. The main users of these services are other service providers, although in an embedded way, other types of users can use them as well.

12.2.3 User's view

The services provided by ISTforCE (Figure 12.4) can be summarised into:

1   Core Infrastructure and Interoperability Services (CIS).
2   Extended Infrastructure Services (MAS, PPS, ECS, TOS, EOS, PDS).
3   Engineering Services (DAS, CCS, RES, SRS, VTLS).

The services in the first and in the second group (i.e. Core and Extended Infrastructure) are conceptually well-defined and comprehensively designed. They provide the backbone of the unique ISTforCE environment and are directly responsible for accomplishing its interoperability. All services of the second group can be external and hence additional arbitrary services can be added. The services in the third group (Engineering Services) are external to the system and are potentially unlimited in number. ISTforCE provides advanced implementation examples featuring the different envisaged methods by which such services can be integrated and made available to the end users.

The user realises the services of the CESP different from their actual arrangement in the software architecture (Figure 12.3). From his point of view, the distributed information space and the different services provided by each exchangeable component can be structured in five distinct services domains as shown on Figure 12.5. These services domains are:

1   Personalised CESP,
2   Core CESP with an Internet-enabled Core Information Server (CIS) and a set of infrastructure services,
3   Local engineering applications,
4   Extended, remote engineering services, and
5   Project data management services.

The domains are elaborated as subsections later. One of their common characteristics is that they have either always one user (1, 3), or are simultaneously

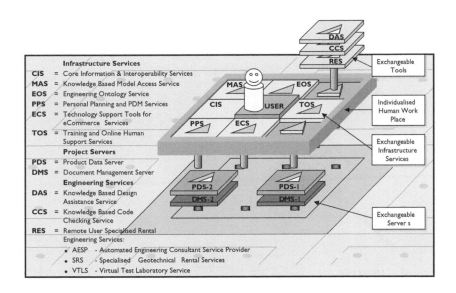

Figure 12.4 The ISTforCE Concurrent Engineering Service Platform (CESP) – The user's gateway to the CE client-server world.

available to many users (2, 4, 5). Figure 12.5 focuses on a single end user, highlighting the main objective of ISTforCE, namely to provide flexible and customisable user-centred design assistance and collaboration services. Therefore if Figure 12.5 is compared to Figure 12.3 the ISTforCE user service layer is sub-divided in domain nos 3, 4 and 5, whereas both infrastructure layers are subsumed to domain no. 2.

The services of the platform, as introduced in Figure 12.4, can be allocated to the five services domains as described later and shown on Figure 12.7.

12.2.3.1 Core CESP

The core of the concurrent engineering services platform (cCESP) covers the

1   core information services (CIS), which manage and control users, projects, services, access rights and billing demands.
2   Extended infrastructure services like

- The Personal Planning Services (PPS) which links the workflows of the projects running in parallel on the platform and enables cross project workflow support to the end users.
- The Model Access Service (MAS) which provides access to one or more product data servers (project spaces) and manages

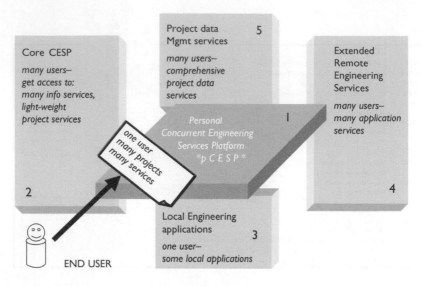

Figure 12.5 The five services domains of the ISTforCE architecture.

local model versions, thereby acting as an information logistics system.

- The engineering ontology service (EOS) which links high-level (human-readable) and technical (machine-interpreted) representations of product model information, thereby enabling easy to use browsing and retrieval of IFC data for engineers that are not familiar with the technical IFC structure and specifications.

- The e-commerce service (ECS) which is not only an e-payment, authentication and certification system but acts also as a negotiation and data interchange tool in order to allow or reject to launch services.

- The training and online support service (TOS) which provides Helpdesk functionality both to the infrastructure and to the external services.

All these services can be provided by third parties due to the common ontology and the standardised interfaces, that is they are not proprietarily fixed components.

12.2.3.2   Personalised CESP

The important end user component, left without sufficient consideration in all mainstream collaboration approaches, is the personalised platform (pCESP) at the user desktop. Typically, in *dotcom* environments, this component is represented only by a standard Web browser, enhanced with relatively simple functional features provided by applets, scripting languages etc. In proprietary frameworks, it is represented by the set of locally installed design tools provided by the same vendor. In contrast, in ISTforCE the pCESP consists of a set of well-defined and structured tools, specifically designed to support collaborative work requirements:

1 A standard Web browser to access the CIS,
2 the generic, configurable Service Launcher to start any remote service,
3 the personal planning service client (PPS/C),
4 the model access service and engineering ontology client (MAS/C), and
5 a set of specific proxy clients, supporting the transparent access to the remote engineering services.

The PPS/C and MAS/C clients do not only act as interfaces to the respective server, which also could have been realised by a standard Web browser but should be capable of starting remote services via the service launcher. The other proxy clients are not mandatory. However they may simplify the access to the corresponding service. Their use is open to the preferences of the particular user. The different kinds of access are illustrated in Figure 12.6, which shows the access diagram of the ISTforCE services. There, the dotted arrows show the alternative business operability supporting access ways, which are explained in more detail further later.

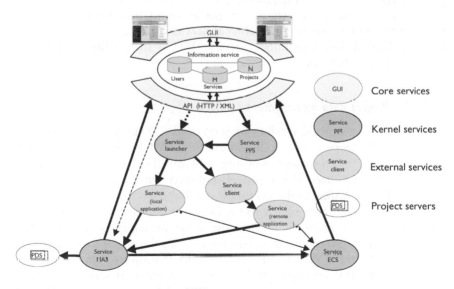

Figure 12.6 Access diagram of the CESP.

12.2.3.3   Local engineering application

Properly registered at the CIS, any local application can easily be integrated in the ISTforCE system and behaves like any other services under one common personalised environment. An example is the Design Assistant Services (DAS) for preliminary structural design and for excavations analysis and design.

12.2.3.4   Extended remote engineering services

The extended remote engineering services (RES) will typically be dispersed over the Internet, hosted by the respective third-party service providers. In Figure 12.7 they are respectively numbered and shown in one 'box' only to emphasise the common approach. By registering them at the core information server and implementing the defined interfaces on the basis of a common high-level ontology specification, backed by more detailed data exchange specifications (see deliverable D9), they can all be made readily visible, selectable and accessible to the individual users of the platform – both directly, through a Web browser, as well as indirectly, through other local or remote applications the user applies. Full added-value of their usage to both the end user and the service providers is achieved by their integration into the environment. They are started from the pCESP with the service launcher. These services are:

- The Code Checking Service (CCS) enabling feasibility checking of the design solution.
- The rental Automated Engineering Consultant Service Provider (AESP) for seismic risk assessment or for special foundation design.
- The Specialised Rental Geotechnical Services (SRS).
- The Virtual Test Laboratory Service (VTLS) for detailed analysis of structural behaviour.

12.2.3.5   Project data management services

They are project-centred services or servers like DMS, PDS or workflow systems. In the scope of ISTforCE, an existing DMS has been enhanced for human-centred multi-project working and an IFC-based Product Data Server (PDS) has been developed. Combined with the MAS a higher functionality according to the ISTforCE requirements has been obtained, in particular coupling more tightly to the infrastructure services. However, its integration in the platform follows the same principle as for the other external services. Moreover, the platform enables the use of more than one different PDS and DMS for different supported projects. Therefore, PDS and DMS can also conceptually be seen as an external service from the point of view of an end-user.

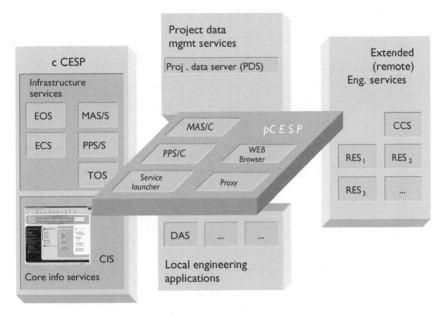

Figure 12.7 Allocation of the components on the five domains of the ISTforCE architecture.

12.2.4   Integration

Integration in the ISTforCE environment is considered along *multiple axes*:

1   Integration of services,
2   Integration of users, and
3   integration of projects, including project tasks and product data.

This is achieved on the basis of a layered set of coherent representations which all use an underlying high-level ontology providing the core concepts of 'User', 'Organisation', 'Actor', 'Role', 'Service', 'Project', 'Model' and the relationships between them. This high-level ontology view on critical system information is explicitly maintained at the Core Information Server and is available to all other components through the XML API, using the HTTP protocol.

12.2.4.1   Integration of services

The basic business idea of ISTforCE is that a lean platform, the CESP is extended by services that are distributed over the Internet and hosted by the respective service providers. The integration is achieved by enabling their

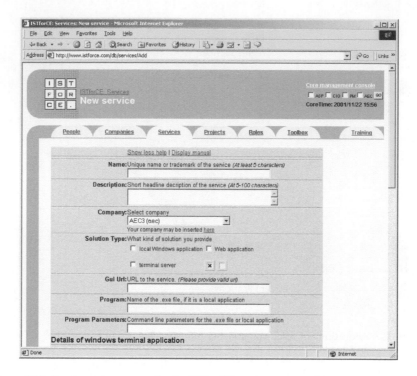

Figure 12.8 Service management in the ISTforCE environment: adding a new service to the platform.

registering in a uniform manner at the Core Information Server (CIS), including all necessary data for their remote invocation, as well as some optional promotion information. *Service registration* is accomplished by the service providers themselves – via a front-end Web interface. Figure 12.8 provides an example.

After a service is registered and respectively described, it can be made available for one or more projects by the CIO who may also restrict its usage to certain users or depending on some other conditions. Registration of the service in the CIS follows a simplified WSDL-like pattern, but also enhanced with attributes related to the user access permissions, payment terms, allowable use of or by other systems etc. Service access is controlled and supported by the 'Service Launcher' which verifies the access rights, certification and ePayment settings for the user and finally launches the service, using an appropriate invocation method retrieved from the service specification at the CIS. Figure 12.9 provides a screenshot of the service selection procedure.

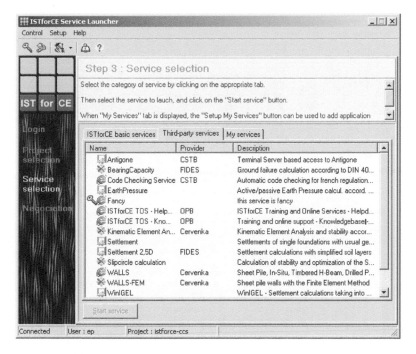

Figure 12.9 Selecting a service by the Service Launcher (only services the user is allowed to access are displayed and offered).

12.2.4.2   Integration of users

An important concept enabling many advanced platform features is the *explicit representation of its users*. For this purpose, the IFC 2× specifications for persons, organisations, actors and roles are taken as baseline. Thus, each user does not only have a login and account, but is also associated with one or more roles for each of his/her current projects. In this way, better supported and more secure access to the project data and the offered services can be warranted.

User management is provided in a similar manner to service management, as illustrated in Figure 12.10. The user information stored at the CIS is extensively used by all other infrastructure services for authentication, but also for appropriate setting of user and services access rights and for eCommerce and ePayment purposes. Also, while normally a user will have the same role and access rights for all of his/her projects, in certain cases these may be different on 'by project' basis. For example, an architect may be only sub-sub-contractor of another architectural office in one project and thus have very restricted access to the building data, whereas in another

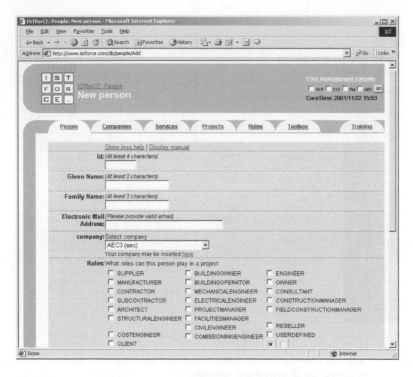

Figure 12.10 User management in the ISTforCE environment: adding a new user and associating him with one or more specific roles.

project he may even play a management role which is usual practice, for example in Germany. Due to the comprehensive user representation, the ISTforCE platform can adequately handle such situations.

12.2.4.3   Integration of projects

At the first glance, this issue does not seem to require any special notice. All state-of-the-art Internet collaboration platforms are capable to host and provide support for multiple projects. However, if the requirement is to support simultaneous cross-project work from the viewpoint of the end user, there is little help that such platforms can currently offer. Problems that need to be adequately tackled include the proper management of personal workflows, access to more than one project data servers, possibly by means of different API specifications, access rights that are separately granted and resolved for each project, various consistency and change management issues, etc.

Because of the greater complexity and the much larger volume of the data that need to be maintained here, multiple project integration in ISTforCE is

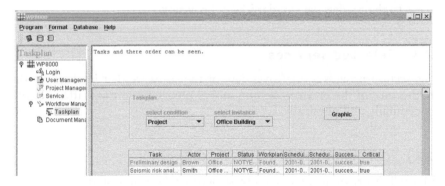

Figure 12.11 Screenshot of the PPS/C enabling user-centric ordering and selection of tasks from multiple projects by appropriate combining and filtering of multiple project workflows.

achieved on *three levels*. High-level meta data describing a project as a whole and the product model involved is stored at the CIS and is managed and used by the other platform services in much the same way as services and users. However, this only provides for a weak degree of consistency and some general queries about the projects. To support the timely performance of project activities, project workflows are stored and managed at the 'Process Planning Server' (PPS/S). Via an appropriate API client applications can retrieve a specific project workflow, a filtered view of the same according to some criteria and, most important, the proper scheduling of the tasks of an individual user for all of his/her current projects. PPS/C is specifically dedicated to support such functionality (see Figure 12.11).

Product data support is even more complex. While, in principle, a similar concept to process management can be envisaged here as well, in practice project data management may be document based, product model based or both. Also, the project data server to use may be prescribed by the client or the main contractor due to the legal responsibility issues involved. Therefore, the approach taken here is inevitably somewhat different. The product data remains completely outsourced at *external* project data servers. They are made available via the specialised *Model Access Server* (MAS/S) which provides a common access point and a uniform API to all other services, hiding the complexity and heterogeneity of the actual server access from the users and applications. To facilitate different data access paradigms, MAS/S uses a plug-in technique enabling the open integration of different access methods. Currently, the implementation includes IFC $2\times$ support based on STEP physical file data exchange, RMI and CORBA. Due to the open architecture, extensions with other methods and model schemas are not difficult to achieve. Additionally, MAS/C provides direct HTTP-based access to the models via standard Web browsers.

12.3   Selected services

This section provids a brief overview of the software that implements the architecture of the previous section and is plugged into the platform, both conceptually and physically.

12.3.1   Core Information Services – CIS

ISTforCE is providing an open infrastructure where almost any component is freely interchangeable. However, there is a need for some very thin infrastructure layer where the tools and services that are plugged in and out are persistently stored. Core information services provide this layer to other parts of ISTforCE – the main users of CIS are other services and they communicate with it using the Internet and the XML format. Information that the CIS is handling is entered via a Web browser therefore CIS is providing a graphical user interface and a shell to connect to the services, people, projects, documents, etc.

The implementation was based on Web services development tool Woda (www.ddatabase.com). To accommodate the need of ISTforCE, an option to provide for machine-readable definitions was incorporated. XML was selected as the documentation format, but the schema of the documentation was proprietary. In principle CIS can be considered an advanced multi-project multi-service directory service for the construction industry (Cerovšek, 2003).

The four types of users shown in Figure 12.1 have each their unique adaptable user interface (see Figure 12.12) represented as side bars.

ISTforCE Core Information Services are implemented as a database managing this information:

- people – People
- companies – Companies they work for
- services – Services they provide/use/like
- projects – Project in which the earlier are involved
- roles – Roles people play in projects
- toolbox – Toolboxes that they have.

The information in those tables can be either accessed interactively, via the GUI using HTML. This is how the CIS are accessed by the humans. Most other ISTforCE services, when they need the core information, access it via the XML based API. In a nutshell, each of the CIS sub-services conforms to Figure 12.13.

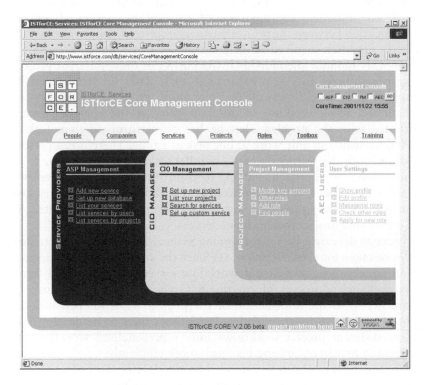

Figure 12.12 Core management console.

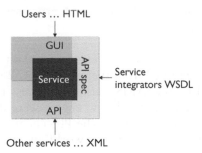

Figure 12.13 Schematic diagram of core services. Interface is through the GUI and API. The latter is specified in a machine readable form.

12.3.2 Personal Data Management and Planning Services – PPS

Planning and constructing a new building require extensive efforts in coordinating processes and data exchange among the different people and organisations involved. By using a workflow management system accessible through the Internet, coordinating efforts will be optimised and the communication between the project partners improved. Due to the fact that engineers are often involved in different projects in parallel, they have to continuously monitor the individual workflows of these different projects. Thus, it is difficult for them to arrange their daily work while ad hoc changes in the workflows of one or more projects regularly occur. Furthermore, when working in different projects the engineer may be confronted with the problem that he has to perform different tasks in parallel, while he can only work at one task at a time. Therefore, a personal workflow, with the restriction of no parallel tasks, has to be created. The PPS introduces an approach enabling to control multi project activities by generating a conflict-free personal workflow without parallel personal tasks and managing multi-project data with the support of the MAS by interacting with distributed project management servers.

In order to develop a Personal Planning System (PPS) several requirements have to be taken into consideration. (1) Since the system has to be integrated into an environment of different workflow servers a standardised format for the exchange of process information has to be used. (2) The engineer has to have access to the information of the project while his personal data has to be controlled independently. (3) Methods have to be developed for merging different project workflows into a personalised workflow that supports the user in organising his work more efficiently.

These functionalities can be provided by the development of a two-layer architecture for process and data management as follows:

- First layer: Containing the project information of one project
- Second layer: Workspace for personal processes and data (Personal Planning Service).

The core of the PPS is the workflow engine based on a relational database management schema adopted from the current IFC $2\times$ schema with some extensions suggested by the 'ARIS' business process model. Furthermore, stored procedures within the workflow engine were developed to control the coherence of processes and data. The PPS client, developed as a JAVA applet, can represent the user's personal workflow in a table as well as in a GANTT Diagram.

By integrating a 'Product Data' and an 'Electronic Data Management System' (PDS and DMS) into the PPS the workflow will not only contain information about time, status and dependency of a task, it furthermore extends the workflow by the view on the data. Thereby, the user will be provided with comprehensive knowledge about the information generated within a project. The access and management of multi-project data is realised with the MAS service described later.

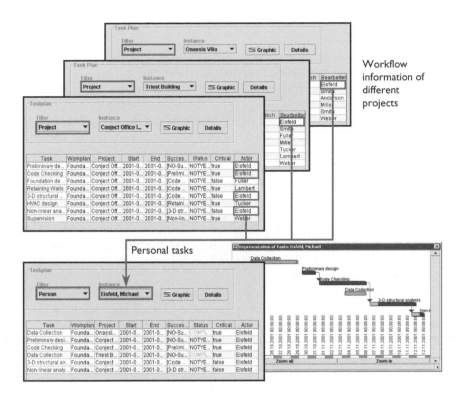

Figure 12.14 Generation of a personal workflow from different projects.

To fulfil the requirements of no parallel tasks in the personal workflow, the critical path as well as the buffer times are calculated for each project. Afterwards all tasks of the user will be filtered out and merged together into a personal workflow. This automatically generated draft workflow can be modified by the user according to his preferences, for instance by distributing the buffer time uniformly over the workflow.

In cases that an overlapping of different tasks were inevitable by the restrictions provided by the project schedules, a delay within the projects will be allowed minimising the sum of all project delays. Mechanisms were developed to support the user rearranging the workflow of the projects in such manner that the influences to dependent tasks are minimised (Figure 12.14).

12.3.3   Product Data Server – PDS

The Product Data Server[2] (PDS) is an external project-centred service that can be plugged in like each other external remote engineering service.

The role of the PDS in ISTforCE is to provide a system that will be responsible of:

- storing data belonging to user projects managed by the server
- keeping track of the various versions of these data
- providing access to product data, either in their whole, or based on a finer granularity
- insuring that only authorised users can permanently save data
- providing services for general users to check-out data, check-in proposals of modified versions of the data
- providing services for administrators to declare user, attributes privileges and rights, etc.

Analysis of common operations done on project data has shown that data access granularity can be defined on a hierarchical relationship based on the containment concept of four entities: (1) IfcProject, (2) IfcSite, (3) IfcBuilding and (4) IfcBuildingStorey. All project data access and queries will use this granularity definition.

The core of the PDS server is built on a component based approach. The overall sequencing of the protocol steps involved in the service of a given command is handled by a core engine, the processing of each command being developed as separate code modules. Using this methodology, it is easy to implement the whole server in an incremental way, adding and testing commands step by step.

The interfaces with the outside is based on extension of FTP commands. Product Model data are exchanged in STEP Physical File format models (ISO 10303, Part21). Structured data not being part of SPF models, such as user profile information or verbose lists, will use a XML format. To give an example of how to use the command set, and to provide a ready to use API, a Java client API is additionally provided. Since the PDS is not intended to be a end-user application, it has no sophisticated user interface. However a simple console is available, so that current status and operations trace can be viewed.

12.3.4 Knowledge-Based Model Access and Engineering Ontology Services – MAS/EOS

The Model access service has to provide human-centred product model services supporting practitioners with additional not directly accessible information and inherent knowledge about their models. Furthermore it is of utmost importance to make product model data and the related product data services easier to access and understand. Therefore the primary objective is to.

- provide customisable retrieval of the additional information and knowledge

- provide user-friendly capabilities for data modification
- enable logistic and concurrent access to the product models of the several projects the engineer is working on at the same time.

Meeting these objectives qualifies the MAS as a central counterpart for product model related ISTforCE services and clients. However, bringing together all these infrastructure services in the MAS paves the way to envisage two additional objectives:

- provide meta model knowledge in terms of an engineering semantics based ontology service to bridge the 'terminological' gap between the IFC model specification and the engineers.
- provide a consistent and extensible framework for server side agents that reason about the product data of the user, utilising the additional information and meta model knowledge of the MAS.

Structuring these infrastructure and advanced services in a coherent approach should yield a particular profit to the ISTforCE CESP.

12.3.5   E-Commerce Services – ECS

The objective is to develop a tool set which supports the civil engineer to hire and remotely pay for the rental of engineering and related services. The tool set will also provide a WWW intelligent negotiation and payment solution within the scope of the engineering service platform. On the other hand, this tool should have completely integrated the certification authority in order to manage the certificates and the secure environment for all processes. Figure 12.15 provides an overview of the supported business process.

12.3.6   Training and Online Human Support
Service – TOS

In every environment questions are arising, which cannot be answered immediately. Within the computer aided planning of projects this can be technical or technological questions and they may considerably impact the project progress. To support the engineer immediately in such situations an extensive online training a service is included in the ISTforCE environment in order to assure and improve the use of the services. Such a help service is also of interest for the service providers. The goal was to develop an intelligent engineering help desk which captures knowledge required and support the use of the services efficiently and support the knowledge transfer gathered from practical experience. It has been implemented as a universally usable service with intelligent, and personalised interface (Figure 12.16).

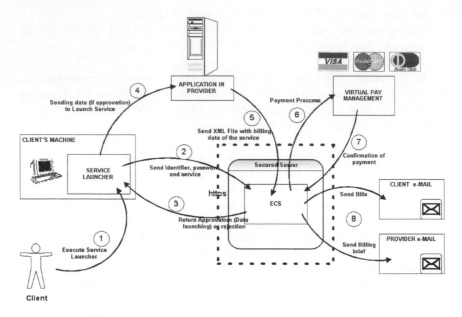

Figure 12.15 ECS processes overview.

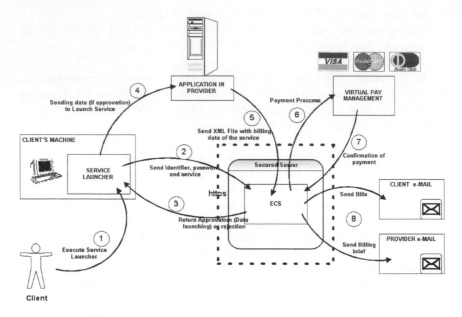

Figure 12.16 Lotus Notes application for support staff.

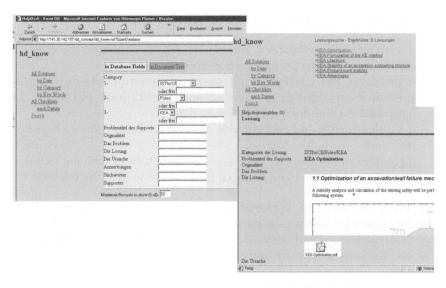

Figure 12.17 Knowledge base – search form, search result and knowledge document.

The TOS is implemented as a Lotus Notes application, running on a Lotus Domino 5.× Server. The Helpdesk staff works fully in the Lotus Notes Environment. They will be informed about each new registered question or change in older ones by e-mail and pop-up notification windows.

A new problem is processed first by a program that tries to find appropriate solutions in the knowledge base. If this is not possible a human resource will be involved. Registered questions can be escalated, cancelled, closed, archived or written to the knowledge base. In the problem database, different views can display the problems sorted by relevant topics like date, employee or supporter. In this environment, the supporter can also access the knowledge base and solve the problems. All interesting and solved problems were written to the Web-based knowledge base. This database also provides training courses for those who want to have a more extensive explanation of the system or the functionality and capabilities of a specific tool or service. Best case examples and training courses, user-oriented and sub-structured, are provided in the knowledge base together with multi-media presentations. In addition, it contains the documentation and tutorials about services and tools. Parts of the knowledge base content are linked from the ISTforCE Core Information Services (Figure 12.17). These links are also displayed in the Service Launcher for fast access to the material.

12.4   Conclusions

The substantial innovations in ISTforCE are that

- it provides a personalised human-centred environment, enhancing current, less flexible project-centred approaches,
- it promotes the concept of an open collaboration platform where various, even third-party services and applications can easily be integrated via standardised XML-based APIs,
- it enables flexible and customisable object level data exchange based on IFC, and
- it provides an infrastructure for online e-business by combining technical capabilities with legal and financial transactions at all system levels.

The key component of ISTforCE is the CE services platform for the collaboration on construction projects as well as a market place for selling construction related services, tools and knowledge. It allows for any construction service or software provider to take part in the new economy.

The key components of the infrastructure for collaborative work are information exchange and communication tools, and the key components of the infrastructure for the providers are service templates, e-Commerce tools and security tools, so that they could concentrate on their core knowledge and not on Internet technology. Below we perform a brief SWOT analysis of the proposed approach.

- *Strengths*. The developed approach, at least as envisaged conceptually, integrates the entire profession in which small and medium companies are in a large majority. It provides them with a new model of doing business and the entire necessary infrastructure. As such it can integrate the fragmented construction profession.
- *Weaknesses*. The prototype is created using tools that allow for rapid prototyping but lack the robustness of the tools with which a professional platform would be built. Currently, we have not addressed issues like security and privacy. The platform also does not provide the definitive answer in relation to the ontology to which the services are supposed to commit. In that respect it takes a rather liberal approach. In fact the minimal commitment is to the rather generic in lean CIS ontology.
- *Opportunities*. Central management of project information should result in a digital archive of previous project. This could enable better reuse of old project data, analysis of the processes as well as synthesising new knowledge about construction. The data could be used to support full life cycle of the structure. To service providers, a common point of entry for all users and a centralised user tracking could lead to better understanding of the users and their needs.

- *Threats*. Companies providing core collaboration services could be tempted into using project data, either discretely or synthetically, to learn about the participants and about the ways construction work is done, and therefore exploiting the implicit knowledge of the construction companies using the services and benefiting from them. An open collaboration platform where many small providers of services and tools can offer these to construction professionals is also a threat to established players in the field, who are interested in exploiting collaboration platforms to extend their monopoly in one segment of the market (e.g. CADD or project planning) over the whole industry. Such portals are also threatened by the general lack of economic soundness on the Internet. In order to establish market shares, dotcoms are offering services nearly for free. Engineering consultants and software authors cannot operate at a similar price. In addition, the ease at which information can be exchanged digitally is likely to cause an information saturation and overload. Designers and planners will be receiving a growing number of messages, files, calls, just because sending out a digital copy of a floor plan is so much easier than drawing out a paper version and mailing it. Therefore, advanced filtering on both sender's and receiver's end will be required as well.

The threats and weaknesses include topics for future research. In the forthcoming inteliGrid project (Interoperability of Virtual Organizations on a Complex Semantic Grid – www.inteliGrid.com) will address the robustness and ontology issues by making the services grid based and committed to a common ontology.

12.5 Acknowledgements

The presented research has been done in the context of 5th Framework IST project ISTforCE, funded partially by the EU. The contribution of the funding agency as well as all the project partners – Obermeyer Planen + Beraten and the Dresden University of Technology (Germany), FIDES (Germany), CSTB (France), GEODECO (Italy), Atlante and APIF (Spain), AEC3 (UK), Cervenka Consulting (Czech Republic) and the University of Ljubljana (Slovenia) – is gratefully acknowledged. The graphics and text of this section borrows from the ISTforCE deliverables that were written by R. Scherer, P. Katranuschkow, Ž. Turk, T. Cerovšek, A. Gehre, M. Magini, R. Juli, J. Cervenka, M. Eisfeld...

12.6 Notes

1 Application programming interface,
2 Recently the IAI has decided to prefer the use of the name project data instead of product data. Here we will mainly use the official ISO 10303 terminology, namely product data.

12.7   References

Bricsnet (2001). Bricsnet Solutions (http://www.bricsnet.com/about/solutions/default.jsp?site=1).

Buzzsaw Inc./Autodesk Inc. (2001). Project Point™ (http://www.buzzsaw.com/content/products_and_services/ProjectPoint/default.asp).

Cerovšek T. (2003). PhD Thesis, University of Ljubljana.

Cervenka J. and Pukl R. (2000). Testing of Building Structures in the Web Towards Virtual Labs, in: Goncalves R., Steiger-Garcao A., Scherer R. J. (eds) Proc. of ECPPM 2000 'Product and Process Modelling in Building and Construction', 25–27 Sept. 2000, Lisbon, Portugal, publ. by Balkema, Rotterdam / Brookfield, ISBN 90 5809 179 1, pp. 97–104.

Citadon Inc. (2001). Collaborative Project Management (http://www.citadon.com/product_center/index.html?pc_collab_proj_mngmnt.html).

Conject A. G. (2001). Conject – Project Area (http://www.conject.de/english/index.htm).

Gehre A. and Katranuschkov P. (2000). Engineering Ontology, ISTforCE Report D5, TU Dresden, Germany, 31p.

Goldberg E. (2001). Architectural Desktop Taps into the Internet, *CADALYST Magazine* 3/2001 (http://209.208.199.147:85/solutions/adt/0301adt/0301adt.htm).

Guarino N. (1998). Formal Ontology and Information Systems, amended version of a paper in Guarino N. (ed.) 'Formal Ontology in Information Systems', Proc. of FOIS'98, 6–8 June 1998, Trento, Italy, publ. by IOS Press, Amsterdam, pp. 3–15.

Hannus M., Karstila K. and Tarandi V. (1995). Requirements on Standardised Building Product Models, in: Scherer R. J. (ed.) Proc. Of ECPPM'94 'Product and Process Modelling in the Building Industry', Dresden, 5–7 Oct. 1994, publ. by Balkema, Rotterdam, pp. 43–51.

IAI (1999). An Introduction to the International Alliance for Interoperability and the Industry Foundation Classes, IAI Publ., Oakton, VA, 21p.

IAI (2001). ST-4 Structural Analysis Model and Steel Constructions (http://www.iai-ev.de/projekte/documents/pdf/IFC_ST4.pdf, http://cib.bau.tu-dresden.de/icss/structural.html).

Katranuschkov P. (2001). Interface Specification for the C/S-based CESP/MAS Project Data Services, ISTforCE RFC-TUD-1, TU Dresden, Germany, 56 p.

Katranuschkov P., Gehre A. and Eisfeld M. (2001). Engineering Ontology, Part II: Formal Representation of the Data Structures, ISTforCE Report D5-2, TU Dresden, Germany, 97 p.

Mangini M. and Protopsaltis B. (2000). E-Commerce: A New Frontier for Engineering Software and Services, in: Goncalves R., Steiger-Garcao A., Scherer R. J. (eds) Proc. of ECPPM 2000 'Product and Process Modelling in Building and Construction', Lisbon, Portugal, 25–27 Sept. 2000, publ. by Balkema, Rotterdam/Brookfield, ISBN 90 5809 179 1, pp. 137–145.

Ouzounis V. (2001). Analysis of Distributed Technologies for the Usage in the Context of Virtual Enterprises. COVE News 1/2001, ISSN 1645 0582 (http://www.uninova.pt/~cove/newsletter.htm).

Scherer R. J. (2000). Towards a Personalised Concurrent Engineering Internet Services Platform, in: Goncalves R., Steiger-Garcao A., Scherer R. J. (eds) Proc.

of ECPPM 2000 'Product and Process Modelling in Building and Construction', Lisbon, Portugal, 25–27 Sept. 2000, publ. by Balkema, Rotterdam/Brookfield, ISBN 90 5809 179 1, pp. 91–96.

Scherer R. J. and Katranuschkov P. (1999). Knowledge-Based Enhancements to Product Data Server Technology for Concurrent Engineering, in: Proc. 5th International Conf. on Concurrent Enterprising, ICE 99, 16–17 March 1999, The Hague, Netherlands.

Turk Ž., Scherer R. J. and Katranuschkov P. (eds) (2000a). Requirements, Specifications, Architecture and Rapid Prototype of CESP, ISTforCE Report D6, 118 p.

Turk Ž., Wasserfuhr R. and Katranuschkov P. (2000b). Environment Modelling for Concurrent Engineering, *Int. J. of Computer Integrated Design and Construction/CIDAC/*2(1), Special Issue on Concurrent Engineering in Construction, SETO, London, UK, pp. 28–36.

Weisberg S. (2001). i-Collaboration – State of the Industry, *CADALYST Magazine* 9/2001 (http://209.208.199.147:85//features/0901icollab/0901icolab.htm).

Weise M., Katranuschkov P. and Scherer R. J. (2000). A Proposed Extension of the IFC Project Model for Structural Systems, in: Goncalves R., Steiger-Garcao A., Scherer R. J. (eds) Proc. of ECPPM 2000 'Product and Process Modelling in Building and Construction', 25–27 Sept. 2000, Lisbon, Portugal, publ. by Balkema, Rotterdam/Brookfield, ISBN 90 5809 179 1, pp. 229–238.

Wix J. and Liebich T. (2001). Industry Foundation Classes IFC 2× (http://www.iai-ev.de/spezifikation/IFC2×/).

(Note: All URL references last accessed in Sept. 2001.)

Chapter 13

Concluding notes

Chimay J. Anumba, Anne-Francoise
Cutting-Decelle and John M. Kamara

13.1 Introduction

This chapter concludes this book and highlights a number of issues relating to the effective implementation of Concurrent Engineering (CE) in construction projects. It starts with a brief summary of the various chapters of the book, emphasises the potential benefits of CE to the construction industry, and discusses the challenges in the practical implementation of CE in construction projects. The last section of the chapter explores some of the future directions in CE and identifies promising research areas.

13.2 Summary

The focus of this book has been on the application of CE concepts to the construction industry. This is in recognition of the huge benefits that the industry stands to reap from the adoption of CE in its project delivery processes. It also supports the trend in industry towards more collaborative working practices.

The introductory part of the book (Chapters 1 and 2) provided a general introduction to CE in construction and its fundamental principles. Chapter 1 presented basic definitions and explored the applicability of basic CE concepts to the construction industry. It also introduced the contents and scope of the book. Chapter 2 focused on the theoretical foundations of CE, and explored its relationship to theories of production.

The second part of the book (Chapters 3–6) addressed practical organisational aspects of CE in the construction industry. The extent to which construction organisations (and the various sectors of the construction supply chain) are ready for the implementation of CE was covered in Chapter 3, which also presented a construction-specific readiness assessment model and its use in the assessment of the construction supply chain. Chapter 4 discussed the importance of capturing the 'voice of the client' within a CE environment and described a tool for doing this within a design context. The applicability of current construction procurement methods in a CE setting was explored in Chapter 5, which saw the adoption of more integrated and collaborative procurement methods in the construction industry as offering an opportunity for the adoption of CE on construction projects. Chapter 6 discussed the importance of process management in construction and, drawing on the Generic Design and Construction Process Protocol, presented the key principles for an improved process for concurrent life cycle design and construction (CLDC).

The third part of the book (Chapters 7–11) covered the technological enablers for CE in construction projects. The focus of Chapter 7 is on the role of ontologies and standards-based approaches in CE. Standardisation efforts discussed include STEP, P-LIB, PSL and IFC. Chapter 8 addressed integrated product and process modelling for CE. Product and process modelling approaches were briefly reviewed and a prototype integrated product and process model, ProMICE, was presented. In Chapter 9, document management in CLDC was covered in detail, including the transition from paper-based documents to electronic and Web-based documents. Chapter 10 used a case study to show how 4D CAD models can facilitate CE by enabling project teams to co-ordinate and plan construction projects more effectively. A telepresence environment that facilitates CE in construction through enabling virtual co-location of project participants and project information was presented in Chapter 11. The provision of support for users within an IT-based CE environment was discussed in Chapter 12 and the approach adopted in a major EU-funded project, ISTforCE, was used to illustrate the key concepts.

13.3 Benefits of CE in construction projects

The chapters in this book have implicitly and/or explicitly outlined the benefits of CE to construction projects and organisations. Some of these benefits derive from similar benefits achieved in other industry sectors while others are based on the anecdotal evidence from construction organisations and project teams that have implemented aspects of CE. Nevertheless, it is useful here to reiterate these benefits:

- Improved quality of facilities relative to cost;
- Reduced duration of capital projects;
- Enhanced efficiency and productivity due to reduction in rework;
- Better co-ordination and management of the construction process;
- Better informed decision making and co-ordination, with decisions taken at the right time and by the right person(s);
- Improved competitiveness of the construction industry relative to other industry sectors;
- Better project definition due to more time provision at the early project stages;

- Improved integration of life-cycle considerations;
- Enhanced collaboration and teamwork between members of the project team;
- More robust information exchange between team members and across the stages in the project delivery process;
- Improved quality of the end product – the constructed facility;
- Greater client satisfaction, given the improved focus on the client's requirements and the delivery of greater value;
- Waste reduction;
- Reduced scope for conflicts and litigation;
- Greater profits for construction companies due to the ability to control more aspects of the project, reducing overall construction time, and improved interaction with designers and other team members;
- Improved safety and 'uptime' for existing operations.

The realisation of the earlier potential benefits depends to a large extent on the effectiveness of CE implementation within the whole construction supply chain rather than in individual firms. However, it is important to note that there is scope for all participants in the construction process to benefit.

13.4   Issues in CE implementation

Construction organisations intending to adopt CE need to address a number of key issues to ensure that they maximise the benefits outlined earlier. It is particularly important that an organisation undertakes a readiness assessment, as discussed in Chapter 3, to ensure that CE implementation is tailored to its specific objectives and business strategy. Some of the main considerations in CE implementation include the following:

- The availability of a robust project development process, which is documented, adaptable, periodically evaluated and facilitates concurrency;
- The existence of an organisational framework and policies that support both individuals and teams, and enables the project development process to be controlled;
- The need for a clear business strategy that outlines an organisation's objectives with regard to interaction with clients and other project team members;
- The agility of an organisation and its capacity to respond quickly to changes in its operating environment;
- The appropriateness of strategies for team formation and operation, including the need to ensure that team members understand their roles and work towards a common purpose;
- Appropriate selection and delegation of authority to team leaders;

- The need for appropriate guidelines for maintaining team discipline;
- The provision of training to enable team members to fulfil their roles and the institution of reward structures that recognise both individual and team achievements;
- Maintaining focus on the client's requirements and having the capacity to respond to any changes that might occur;
- The institution of appropriate procedures and policies for quality assurance;
- The development of designs that are flexible, robust and informed by the client's requirements;
- The availability of appropriate technologies to facilitate information exchange and knowledge sharing;
- The use of an integrated project model and systems that facilitate integration between members of a project team;
- Use of common hardware and software platforms to ensure the seamless exchange of information on projects;
- Use of standard and proven information and communications technologies.

The earlier list is not exhaustive but includes the majority of issues that need to be considered. It should also be pointed out that there are many barriers to the uptake of CE in construction, and consideration needs to be given to overcoming these to ensure successful CE implementation. In this regard, some of the main barriers that need to be overcome include:

- The fragmentation and traditional adversarial relationships between team members;
- The lack of trust between team members;
- The lack of a recognised stakeholder for overall process improvements;
- Traditional adherence (usually by government bodies) to a 'lowest bidder' model of tendering rather than best value;
- Conservative nature of the construction industry;
- Low levels of awareness and understanding of the principles and benefits of CE.

These barriers can be addressed in a variety of ways but by far the most promising approaches include the following:

- Improvements in education and training for both new entrants and established practitioners in the construction industry;
- Provision of incentives for collaborative working;
- The use of demonstration projects with innovative clients to showcase the benefits of the CE approach;
- Changes in government regulations, particularly with regard to competitive bidding;

- The adoption of established information and communications technologies (e.g. groupware, 3D modelling, Web-based project collaboration systems, etc.) that facilitate collaborative working;
- The establishment of strategic alliances and partnerships.

13.5   Future directions

CE is sometimes referred to as a 'container philosophy' in terms of the fact that it has a number of core principles but is not prescriptive about how these are achieved. This means that several other initiatives such as lean construction, value engineering, quality management and partnering can all fit into the CE agenda. Thus, it is impossible to predict, with any credibility, the direction that CE will take in the future. This section will, therefore, simply draw on a number of industry trends and expert surveys to outline some of the issues that will have an impact in shaping the future:

- The issue of trust is central to collaborative and CE in construction. This is now being increasingly recognised and a number of research projects have been undertaken. However, there is still scope for the development of effective models and frameworks to facilitate trust building.
- The growth of knowledge management in industrial practice will have an impact on the implementation of CE in construction, as it will enable CE best practice to be propagated. In this regard, appropriate tools and techniques need to be developed to facilitate collaborative learning within construction project teams.
- The advent of the Semantic Web is expected to facilitate more knowledge sharing and collaborative working but the difficulties in construction remain. For example, the development of an appropriate construction ontology remains elusive. Furthermore, there are no guidelines or tools for project information management in a Semantic Web environment.
- Action research is required on those projects that adopt (aspects of) CE, so that the lessons learnt and critical success factors can be captured and shared.
- Considerable changes are required in construction project delivery processes if the full benefits of CE are to be realised. So far, there are very few models of how these processes should be re-engineered within a CE environment.
- In addition to process change, there is also a requirement for changes at organisational level. Further studies are necessary to establish the most appropriate ways to institute changes at individual, team and organisational levels, and (crucially) at the interfaces between these.

- CE adoption will redefine the *modus operandi* of construction supply chains. More work is needed to understand how best to integrate CE principles into supply chain management.
- Sceptics will remain unconvinced until the advantages of CE over conventional approaches can be quantifiably demonstrated. This calls for the development of appropriate metrics for the evaluation of CE performance, although it must be recognised that not all benefits can be quantified.

It is evident from the contents of this book that CE has much to offer construction-sector organisations. The complexity associated with the delivery of construction projects by a transient project team made up of individuals/teams from a variety of organisations makes the implementation of CE challenging. However, this also makes the successful implementation of CE in construction projects highly rewarding for all members of the project team. Construction organisations need to establish CE implementation plans at both individual organisation level and project organisation level (i.e. involving the whole supply chain), if the industry is to reap the full benefits of CE adoption.

Index